# WELDING SKILLS AND TECHNIQUES

# WELDING SKILLS AND TECHNIQUES

Robert P. Schmidt

 Reston Publishing Company
A Prentice-Hall Company
Reston, Virginia 22090

**Library of Congress Cataloging in Publication Data**

Schmidt, Robert P.
  Welding skills and techniques.

  Includes index.
  1. Welding.  I. Title.
TS227.S314        671.5′2        81-19263
ISBN 0-8359-8611-X        AACR2

© 1982 by Reston Publishing Company, Inc.
*A Prentice-Hall Company*
Reston, Virginia 22090

10  9  8  7  6  5  4  3  2

Printed in the United States of America

# CONTENTS

# PREFACE

This book is the result of my determination, with the encouragement of others, to put together all of the materials gathered and used during many years of working and teaching in the welding field.

Before writing the text, I reviewed many books on welding and found them lacking in some areas or that their information was presented in the wrong manner. Many welding books were written for bench welding, whereas *Welding Skills and Techniques* is slanted toward on-the-job or field repair welding.

This book was written for the purpose of helping the general welder or repairman, those who want to learn welding, and those who wish to become professional welders.

The information presented is general in nature and can be understood by the average student or layman. In the acetylene and electric arc, or stick electrode, section, the automobile body is mentioned frequently when discussing repair welding. The purpose in using the automobile body is that, owing to its popularity, one can more easily follow the text when learning about types of joints, welding positions, or problems that may be encountered when welding. Moreover, the MIG Metallic Inert Gas Welding process, or gas metallic arc welding (GMAW), will be used more and more in auto repairing as the automotive industry continues to increase its use of alloy steels.

Therefore, this book is directed to the many persons who are involved in the welding repair trade.

# ACKNOWLEDGMENTS

This book became a reality because of the help of many people, all of to whom I am deeply indebted. I would like to express my sincere gratitude and appreciation to my wife, Estelle, for her many hours of work in helping me prepare the text; and to my children, especially my daughters Jeanne and Lisa, for their assistance in the art work.

Many thanks to my close friend, Mr. Nick Contino of Prentice Hall, Inc., who encouraged me to start this project and who introduced me to Reston Publishing Company. Thanks to David Dusthimer, editor of Reston Publishing Co., for his assistance and for giving me the opportunity to finish this book. Thanks also to the Reverend Stuart Suydam of the Callicoon Center Dutch Reform Church, who helped me with the photographs.

Thanks to the administration and staff of the Sullivan County Board of Cooperative Educational Services, Occupational Division; and the Sullivan County Community College, Continuing Education Division; for their help and assistance. Also, thanks to all my vocational teachers, business associates, and friends for their support.

Last, but not least, thanks to the many manufacturers who supplied me with the necessary materials.

# WELDING SKILLS
# AND TECHNIQUES

# OXY-ACETYLENE WELDING AND CUTTING

## chapter 1

In large scale production or manufacturing, the oxy-acetylene welding process has been generally replaced by newer fusion and welding methods. In the repair and maintenance field, however, the oxy-acetylene welding and cutting torch is still one of the most important pieces of equipment. It is a "must" for repairing automobile bodies and for such mechanical repairs as on damaged exhaust systems or for freeing rusted bolts and nuts (see Figure 1-1). Use of the oxy-acetylene torch ranges from repairing or salvaging large maintenance equipment to repairing smaller items, such as kitchen chairs and other household articles.

The cutting torch is also used for dismantling huge steel structures, cutting up scrap auto-mobiles, or in any place where steel or iron must be severed.

The oxy-acetylene welding equipment is used for welding, cutting, heating, bending, and soldering (see Figure 1-2). With a wide assortment of different sized welding and cutting torches and a good selection of tips, the torch can supply sufficient heat for various thicknesses of metals. Many things can be repaired if the repairman is a good welder.

It is very important for anyone entering the auto body field to learn and become familiar with the oxy-acetylene welding torch and its operation.

Misusing the oxy-acetylene torch can cause injuries, gas explosions, fires, and other dangers.

1

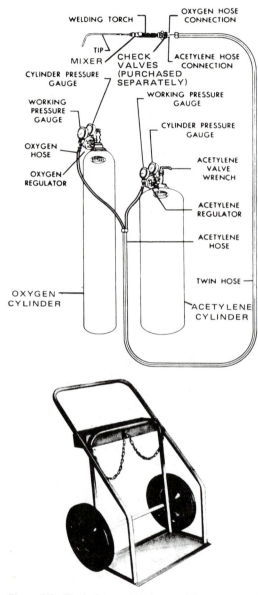

Figure 1-1 Typical oxy-acetylene welding setup and cart. (*Courtesy of Airco.*)

However, with reasonable care and knowledge, the torch is as safe as any hand tool. Carelessness in handling the equipment and cylinders, and failure to exercise precaution when using the torch around flammables, can make the torch a very dangerous tool. Most injuries and damage are caused by human error rather than equipment failures.

## HISTORY OR DEVELOPMENT OF ACETYLENE AND OXYGEN

Acetylene gas was discovered by accident around 1840, but it was not until about 1890 that acetylene gas was produced from calcium carbide.

The availability of acetylene gas for commercial use, along with oxygen, depended on devising a better method for producing calcium carbide and obtaining oxygen. The use of the electric furnace with its high temperatures led the way for increased production of calcium carbide. Around the same time, a method was discovered to liquify air as a means of obtaining oxygen.

At the turn of the century, the first practical blowpipe was put into use to control the oxygen-acetylene flame. As methods improved for the production of acetylene and oxygen, they were made available in economical form. Gradually, a suitable container or cylinder was developed for oxygen, and the idea of compress-

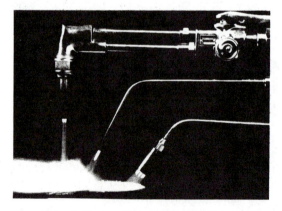

Figure 1-2 Cutting, welding and heating. Some of the uses of the oxy-acetylene equipment. (*Courtesy of Victor Welding and Cutting Division.*)

ing oxygen for storage in the cylinders became reality.

During the early part of this century, it was discovered that the acetylene flame gave off a brilliant light and it could be used for lighting buildings, streets, and rural homes. Some automobiles used acetylene lamps for light. However, it soon became evident that acetylene gas could not be stored in the same type of cylinders as oxygen.

The oxy-acetylene flame produces a very high temperature; in fact, it is the hottest flame known, approximately 6300°F (3482°C). It is capable of melting all commercial metals; fortunately, it can be localized and controlled. The flame does not affect the metal melted, and the molten metal is protected from atmospheric contamination or from oxygen and nitrogen attacking the molten metal.

As new techniques for oxy-acetylene welding were discovered, the acetylene and oxygen became available in greater supply. This led to the rapid growth of the oxy-acetylene welding business.

In the early years of oxy-acetylene welding, many shops produced their own acetylene gas by means of a low pressure acetylene generator. Eventually, the present acetylene cylinder became available in greater supply and replaced generators.

## ACETYLENE

Acetylene gas is made by mixing calcium carbide with water. Calcium carbide itself is made by fusing limestone and coke at a very high temperature. After it has cooled sufficiently, the fused block is crushed into granules or crystals.

The acetylene gas is made in an acetylene generator by mixing or dropping a measured amount of calcium carbide crystals or granules into water, which starts a chemical reaction.

The calcium combines with the oxygen from the water to form a strong solution of lime which settles to the bottom of the generator. This leaves two parts or atoms of hydrogen from the water and two parts or atoms of carbon from the calcium carbide. These unite to make acetylene—$C_2H_2$. To put it in simple language, the carbon (C) and the hydrogen (H) do not have a great liking for each other, so the acetylene gas becomes unstable when compressed above 15 psi (103.5 kPa) in a free state. The two elements can disassociate or come apart at the slightest aggravation such as a shock or application of heat. When acetylene gas disassociates, it gives off sufficient heat to ignite. This is why acetylene must be stored in a special tank or cylinder (which will be discussed later). Acetylene is colorless and has a very strong odor like garlic.

## OXYGEN

There are numerous ways to make or produce oxygen but some of them involve expensive processes. The most common methods are chemical electrolysis and liquifying of air. Most of the oxygen is produced by liquifying air because it is less expensive than the chemical electrolysis method.

The chemical electrolysis method involves passing a current through water that contains another chemical. The oxygen is given off at one terminal plate and hydrogen at the other plate.

To appreciate the liquifying of air method, we must remember that the air we breath, or the atmosphere, contains about 20% oxygen and about 80% nitrogen with the rest being small percentages of argon, helium, carbon dioxde, and water vapor.

The process of liquifying air is rather complicated and involves many different stages. However, a simplified explanation follows. First the air is forced into a large container, then cleaned, filtered, and compressed. The air is cooled then expanded to atmospheric pressure. The change from high pressures under very low temperatures to a lower pressure causes the air to liquify—the air changes from a gas to a liquid. The liquid air is an extremely cold mixture. The oxygen and nitrogen can now be separated because of the difference in their boiling points. Oxygen has a boiling point of $-297.2°F$ ($297.2°F$ below zero, which is $-182.9°C$). The nitrogen boiling point is $-319°F$ ($-195°C$). The nitrogen, having a lower boiling point, will evaporate first leaving the oxygen. The oxygen, about 99.5% pure, is heated to a gas and pumped into storage tanks.

## STORAGE OF THE GASES IN CYLINDERS

### OXYGEN CYLINDERS

Because of the high pressure under which the oxygen is stored in the oxygen cylinder, the cylinders must pass rigid requirements of the Department of Transportation.

The oxygen cylinders are made of a seamless high carbon steel no less than 3/8" (9.52 mm) thick. The normal full tank of oxygen contains 2200 pounds per square inch, psi (15,170 kPa) at 70°F (21°C). The oxygen will expand when heated and contract when cooled so that the pressure may vary depending on the temperature.

The oxygen cylinders are tested periodically to a pressure over 3300 psi (22,750 kPa). The inside walls may deteriorate over a period of time and must be checked for any weakness. This is why most companies will not sell their cylinders but rent or lease them out to insure the safety of the users.

A malleable iron neck is shrunk in at the top of the cylinder and the tank valve is screwed into the neck (see Figure 1-3). Also, the neck

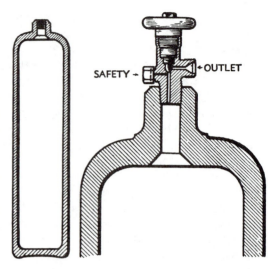

Figure 1-3 Oxygen cylinder and valve. (*Courtesy of Modern Engineering Co.*)

is threaded on the outside to permit a protective cap to be attached. The valve cap protects the valve if the cylinder is accidentally knocked over and while the cylinders are being transported. If the valve is accidentally broken off when the cylinder pressure is about 2000 psi (13,780 kPa), the very high pressure of the escaping oxygen would burn anything it came in contact with and the cylinder would be jet propelled.

NOTE: Never leave a cylinder, either empty or full, to stand by itself without adequate support.

Store in an upright position and away from heat. Always keep the cylinder valve closed when empty to avoid having foreign particles enter the cylinder. Keep oil or grease away from the cylinder—especially the valve. Do not strike an electric arc on the cylinder because it may cause it to rupture.

The cylinder valve is made of forged brass and made to withstand high pressure. The valve is called a back-seated or double-seated valve (see Figure 1-4). The back-seated part of the valve seals off the thread to prevent oxygen from escaping. The cylinder valve must be

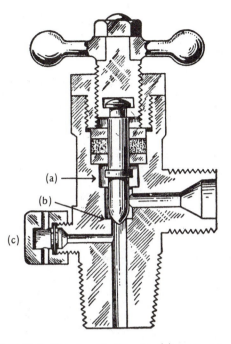

Figure 1-4 Oxygen cylinder valve; (a) upper seat seals threads; (b) lower seat closes cylinder; (c) safety disc. (*Courtesy of Linde Co., Division of Union Carbide.*)

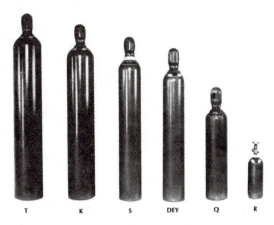

Figure 1-5 Since gases have a relatively low density, a given volume of gas at atmospheric pressure can be considerably reduced by compressing it into a cylinder under greater pressure. These cylinders must be constructed to withstand the high pressures involved. LINDE high-pressure gas cylinders not only meet DOT specifications but must also meet additional rigid specifications set up by Union Carbide. Quality control, from chemical analysis of molten steel to final testing, is rigidly controlled. A variety of cylinder sizes is available for all industrial gases. (*Courtesy of Linde Co., Division of Union Carbide.*)

opened all the way in order to seal the threads. Partially opening the valve will waste oxygen. Oxygen valves have right-handed threads.

Each valve is equipped with a safety pressure disc that will burst when the pressure gets too high (caused by excessive heat). This will prevent a rupture of the cylinder.

The most common sizes of oxygen cylinders used are the 122 cu. ft. (3454.1 liters) and the 244 cu. ft. (6909.3 liters) with some of smaller and larger capacities (see Figures 1-5, 1-6, and 1-7).

NOTE: Never use oil or grease on a cylinder or any part of the welding set-up at any time because a serious fire or explosion could result. The cylinders are painted different colors for different gases; however, the colors vary with each distributor or manufacturer.

It is not advisable to use the oxygen cylinder

after the pressure is down to 100 psi (690 kPa) welding or 200 psi (1380 kPa) for cutting because a serious backfire could result if the oxygen pressure were less than the acetylene.

## ACETYLENE CYLINDERS

Acetylene, because of its nature, must be stored in a different type of cylinder than that used for oxygen. Acetylene cannot be stored in a hollow cylinder at a pressure over 15 psi (103.5 kPa) in a gaseous form because of its explosive nature. In order to store acetylene in a cylinder at a pressure up to 250 psi (1720 kPa), the cylinders are filled with a porous material such as Fuller's Earth or wood pulp, depending on the manufacturer (see Figure 1-8). The filler material must be approved by the Department of Transportation in order for the cylinder to be shipped between states. Next, acetone is

| GAS | Cylinder Style | Contents (cf) | Full Cyl. Pressure at 70° F. (psi) | Height incl. Cap (in) | Outside Diameter (in) | Approximate Weight | | Cylinder Valve Outlet Connection CGA No. |
|---|---|---|---|---|---|---|---|---|
| | | | | | | Full (lb) | Empty (lb) | |
| OXYGEN | T | 330 | 2.640 | 60 | 9-1/4 | 172 | 146 | 540 |
| | K, KL | 244 | 2.200 | 56 | 9 | 153 | 133 | 540 |
| | LK | 244 | 2.200 | 56 | 8-15/16 | 134 | 114 | 540 |
| | D | 122 | 2.200 | 48 | 7-1/2 | 126 | 116 | 540 |
| | E, Y | 122 | 2.200 | 48 | 7 | 92 | 82 | 540 |
| | Q | 80 | 2.200 | 35 | 7-1/8 | 70 | 65 | 540 |
| | XL | 70 | 2.200 | 41 | 6 | 54 | 49 | 540 |
| | S | 150 | 2.200 | 51 | 7-3/8 | 92 | 80 | 540 |
| | R | 20 | 2.200 | 19 | 5-3/16 | 13-1/2 | 12 | 540 |
| ACETYLENE | WTL | 390 | 250 | 44-3/8 | 12-1/2 | 207 | 180 | 510 |
| | WK | 304 | 250 | 42 | 12-7/8 | 245 | 223 | 510 |
| | WSL | 130 | 250 | 33-1/2 | 8-1/2 | 78 | 69 | 510 |
| | WS | 130 | 250 | 35-1/2 | 8-1/2 | 79 | 70 | 510 |
| | WC | 111 | 250 | 37-1/2 | 8-1/2 | 94 | 87 | 510 |
| | WQ | 60 | 250 | 24-1/4 | 7-5/8 | 56 | 52 | 510 |
| | B | 40 | 250 | 23 | 6-1/4 | 26 | 23-1/2 | 520 |
| | MC | 10 | 250 | 14 | 4 | 8 | 7 | 200 |

**Figure 1-6** Cylinder and valve data. (*Courtesy of Linde Co., Division of Union Carbide.*)

| Spec'n | OXYGEN | | ACETYLENE | | MAPP GAS | | PROPANE | | NATURAL GAS | |
|---|---|---|---|---|---|---|---|---|---|---|
| | U.S. | Metric | U.S. | Metric | U.S. | Metric | U.S. | Metric | U.S. | Metric |
| Diameter | 9 in. | .23 m | 12 in. | .31 m | 12 in. | .31 m | 15 in. | .38 m | 9 in. | .23 m |
| Height | 55 in. | 1.4 m | 40 in. | 1.02 m | 44 in. | 1.12 m | 50 in. | 1.27 m | 55 in. | 1.4 m |
| Capacity | 244 cu ft | 6.9 m$^3$ | 250 cu ft | 7.1 m$^3$ | 600 cu ft* | 17 m$^3$* | 950 cu ft* | 27 m$^3$* | 220 cu ft | 6.2 m$^3$ |
| — | — | — | — | — | 70 lbs* | 32 kg* | 115 lbs* | 52 kg* | | |
| @ Temp. | 70°F | 21°C | 70°F | 21°C | 60°F | 16°C | 70°F | 21°C | 70°F | 21°C |
| @ Pressure: | 2200 psig | 15,200 kPa | 250 psig | 1725 kPa | 94 psig** | 650 kPa** | 125 psig | 860 kPa | 2200 psig | 15,200 kPa |

\* Purchased by Weight
\*\* Vapor pressure at 70°F or 21°C.

**Figure 1-7** Cylinder data. (*Courtesy of Airco.*)

added and is absorbed by the porous material. Acetylene is then pumped into the cylinder to a pressure of about 250 psi (1720 kPa). The acetone readily absorbs many times its own weight of acetylene. In addition the porous material prevents pockets of acetylene gas from accumulating. This method is the safest way to store acetylene. The cylinders are constructed in accordance with the Department of Transportation requirements. Some cylinders have concave bottoms and two safety plugs that will melt out at around 220°F (104°C) to relieve the pressure in case of excessive heat (see Figure 1-9).

Two types of valves are used for acetylene cylinders. One type uses a hand wheel; the other type uses a square shank and key.

NOTE: When the cylinder is being used, the key must be kept on the cylinder at all times to permit closing the valve in case of emergency.

The valves are of a simple construction because of the lower cylinder pressure as compared to the pressure in the oxygen cylinder. The tops of the cylinders vary between different manufacturers. Some have recessed tops to protect the valve, while others have tops shaped similar to the oxygen cylinder, but the cylinder is wider.

The acetylene cylinder has left-handed threads on the cylinder outlets, to avoid putting the wrong regulator on. The oxygen regulator is made to withstand a much higher pressure as compared to the acetylene regulator.

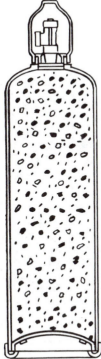

Figure 1-8 Cross section view of acetylene cylinder with porous filling material. (*Courtesy of Victor Welding and Cutting Division.*)

Head ring helps protect cylinder valve from damage

- Sturdy steel cylinder body

Fuse plugs control the release of acetylene contents should temperature outside or inside cylinder exceed 212° F (100° C) for protection in case of a fire

Foot ring to protect cylinder from moisture and corrosion

Figure 1-9 Cylinder construction. (*Courtesy of Linde Co., Division of Union Carbide.*)

*CAUTION: Never attempt to transfer acetylene from one cylinder to another or to another container.*

NOTE: Acetylene is sold by weight, not pressure. This is determined by weighing the cylinder after it is filled and subtracting the weight of the empty cylinder. The difference is multiplied by 14.5 (14.5 cubic feet equal 1 pound, 410.5 liters equal 0.453 kilograms).

Due to the nature of acetylene, the gas is given off by the acetone at a slower rate at lower temperatures. It may be difficult to get sufficient heat from a large welding or cutting tip if the cylinder is exposed to below freezing temperatures.

ALWAYS store acetylene cylinders in an upright position as acetone may enter the valves

and gauges. Store away from heat. The acetylene cylinders are available in many different sizes (see Figure 1-10). The 110- and 130-cu. ft. (3114 and 3770 liter) cylinders are used in the average shops. The 300-cu. ft. (8495 liter) cylinders are used in larger shops, generally for multi-torch stations or on manifold systems (see Figure 1-11).

## MAPP GAS

Another fuel gas that can be used for welding, cutting, brazing, and heating is called MAPP gas. MAPP is a trademark of Airco, Inc., and is a mixture of stabilized methacetylene and propadiene.

The oxy-MAPP flames have a high flame temperature of 5301°F (2915°C), while the oxy-acetylene high flame temperature is approximately 6000°F (3304°C). MAPP gas may be used at about 90 psi (621 kPa) regulator pressure while acetylene is limited to 15 psi (103.5 kPa). MAPP gas is shipped in a liquid state in cylinders from 7 to 1,800 lbs. (3.17 kg to 816.4 kg). MAPP is insensitive to shock and

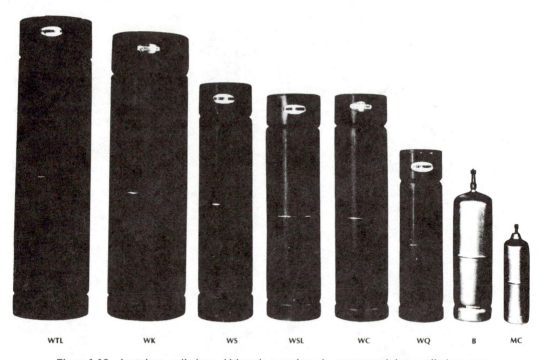

WTL          WK          WS          WSL          WC          WQ          B          MC

**Figure 1-10** Acetylene cylinders. Although acetylene is compressed into cylinders at a lower pressure than other industrial gases, LINDE acetylene cylinders are manufactured to the same high standards as those used for other gases. Acetylene in commercial cylinders is supplied dissolved in acetone since acetone will hold more than 400 times its own volume of dissolved acetylene at 70F and 250 psi full cylinder pressure. Union Carbide has developed a special filler to retain the acetylene-acetone solution. This exclusive filler, with 89 percent porosity, provides reduced cylinder weight, increased cylinder capacity and improved charging and discharging rates. (*Courtesy of Linde Co., Division of Union Carbide.*)

**Figure 1-11** Portable acetylene manifold. (*Courtesy of Airco.*)

the cylinders will not detonate when dented or dropped.

MAPP gas can be used with the standard acetylene regulators with the cylinders having the same type threads as the acetylene cylinders. The same welding torch and cutting assemblies can be used for MAPP gas with the change in tips. There are torches made especially for MAPP gas; they are available at the MAPP distributors.

The oxy-MAPP flame is wider and is excellent for welding of heavy plate and for brazing (see Figure 1-12). If fine, pinpoint welding is required, acetylene is better than MAPP gas. MAPP gas

**Figure 1-12** Brazing with MAPP gas. (*Courtesy of Airco.*)

will cut faster than acetylene and can be used at greater underwater depths because of the higher available regulated pressure.

In the weight and safe handling comparison, a filled acetylene cylinder weighs about 240 lbs. (108.8 kg) as compared to a cylinder of MAPP weighing 120 lbs. (54.5 kg). The acetylene cylinders are either rented (demurrage) or contracted, but the MAPP cylinders are generally customer owned, saving on the rental costs. In the following sections, acetylene gas will be discussed because, at the present time, acetylene is the most commonly used for welding.

## REGULATORS

In the previous discussion of the cylinder, it was noted that the pressure of the oxygen cylinder is about 2200 psi (16,170 kPa), while the acetylene cylinder is about 250 psi (1720 kPa). This pressure is too high to be used for welding. Therefore, some means must be used to reduce the pressure so that it is suitable for welding or cutting. In addition, a constant pressure must be maintained for welding. Regulators are used to perform these two important functions.

The regulator will maintain a constant pressure at the torch even though the pressure of the cylinder decreases as the gas is consumed. On occasion, some regulators may have to be readjusted as the cylinders become nearly empty.

For example, with 2200 psi (15,170 kPa) in the cylinder and 5 psi (34 kPa) of oxygen needed at the torch, the pressure will remain constant. After a prolonged period of welding or cutting, the cylinder pressure may drop to 500 psi (3450 kPa), but the torch pressure will remain constant at 5 psi (34 kPa). Even if a large cutting tip is used where the required pressure of oxygen may be 60 psi (414 kPa) at the torch, the pressure must remain constant.

Regulators are high quality, sensitive instruments made for accurate regulation of pressure but rugged enough for shop handling. The regulators for both acetylene and oxygen operate on the same principle. Each regulator has two gauges, one that indicates the approximate cylinder pressure and the other for the approximate amount of regulator pressure that will be delivered to the torch (see Figure 1-13).

NOTE: As stated, one gauge indicates the approximate amount of regulated pressure or cylinder pressure. This means that not all regulators will indicate the exact pressure; the gauges may have become worn or damaged by handling.

One indication of the gauges not being accurate: the acetylene regulator is set at 5 psi (34 kPa) and the torch will not light. However, by setting the regulator at 10 psi (68 kPa), the torch will light. This means that the regulator gauge is off or out of calibration. Also, when the cylinders are shut off and the gas drained from the hoses, the low pressure gauge may indicate pressure.

Opening the cylinder valves too fast, especially the oxygen cylinder, is one of the chief causes of damage to the gauges. Most gauges use a Bourdon tube (a thin curved brass tube) which tends to straighten out as the pressure increases.

## OPERATION OF THE REGULATOR

The basic operation of the regulator is fairly simple. After the cylinder valve is opened, the regulator adjusting screw is turned inward, or clockwise, to start the flow of gas to the torch (see Figure 1-14). The adjusting screw compresses the pressure adjusting spring. The spring in turn exerts pressure on the diaphragm, which moves downward moving the valve or needle off the valve seat permitting the gas to flow from the high pressure side of the regulator to the low pressure side and to the torch. When the torch is shut off, the pressure will build up in the low pressure side and, aided by the valve return spring, will push the diaphragm upward, shutting off the flow of gas.

The two forces, the adjusting spring tension and the pressure of the gas, when balanced, will provide an even flow of gas to the torch.

On occasion, the gauge will indicate that the pressure is increasing after the torch has been shut off. This is caused by a worn out seat or needle or could be caused by dirt on the seat. If the pressure becomes excessive, it could

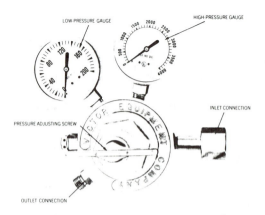

**Figure 1-13** Exploded view of a regulator (exterior view). (*Courtesy of Victor Welding and Cutting*.)

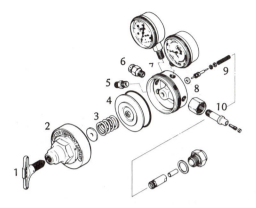

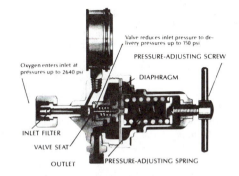

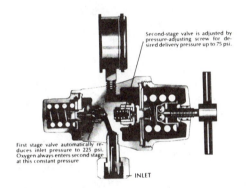

**Figure 1-14** Exploded view of a regulator (cross section). 1. adjusting screw; 2. bonnet or cap; 3. adjusting spring; 4. diaphragm; 5. low pressure outlet; 6. safety valve or disc; 7. body; 8. needle and seat; 9. counter spring; 10. high pressure inlet. (*Courtesy of Victor Welding and Cutting.*)

**Figure 1-15** Cross section of a single-stage oxygen regulator. (*Courtesy of Linde Co., Division of Union Carbide.*)

**Figure 1-16** Cross section of two-stage oxygen regulator. (*Courtesy of Linde Co., Division of Union Carbide.*)

rupture the diaphragm or possibly blow the Bourdon tube in the gauge. This is generally called creep, and the regulator should not be used until it is repaired.

The regulator mentioned above is called the single-stage regulator and is most commonly used in small shops (see Figure 1-15).

## *TWO-STAGE REGULATOR*

The two-stage regulator regulates the gas twice and is more constant as well as more accurate than the single-stage regulator where large volumes of gas or gases are being used at one time (see Figure 1-16).

In the first stage, which is preset, the oxygen is reduced from approximately 2200 psi (15,170 kPa) to about 225 to 300 psi (1555 to 2070 kPa) depending on the make of the regulator. The second stage is similar to a single-stage regulator, which reduces the pressure from 225 to 300 psi (1555 to 2070 kPa) to the desired working pressure.

The acetylene regulator, the preset first stage, reduces the pressure to about 50 to 75 psi (345 to 517 kPa) depending on the make of the regulator. Most of the oxygen regulators

are equipped with a bursting disc. This relieves the pressure inside the regulator.

The outlet fittings on the oxygen regulators have right-handed threads. The acetylene outlet fittings have left-handed threads. The fittings are grooved or notched to indicate fuel gas fittings.

## OTHER EQUIPMENT

### *HOSES AND FITTINGS*

The hoses are used to carry the gases from the regulator to the torch. The hoses must be of a special construction and made of a non-porous material to withstand pressure, to take

## TWIN FITTED HOSE ASSEMBLIES— TYPE VD, GRADE RM

- consist of oxygen and fuel gas hoses of identical construction, with heavy braid reinforcement and continuously connected along sides—oxygen hose is green, fuel gas hose is red

- hoses have smooth, flame- and oil-resistant neoprene covers and non-oil resistant inner rubber tubes

- meet standards of Compressed Gas Association, Rubber Manufacturers Association, Association of American Railroads and National Fire Protection Association

### ORDERING INFORMATION

| | Length | | Connection |
| | ft | m | Sizes |
| --- | --- | --- | --- |
| 3/16 in. Whip Hose Assembly | 10 | 3 | B-B |
| 3/16 in. Twin Fitted Hose Assembly | 12-1/2 | 3.8 | A-A<br>A-B<br>B-B |
| | 25 | 7.6 | A-A<br>A-B<br>B-B |
| | 50 | 15 | B-B |
| 1/4 in. Twin Fitted Hose Assembly | 25<br>50<br>100 | 7.6<br>15<br>30 | B-B<br>B-B<br>B-B |
| 5/16 in. Twin Fitted Hose Assembly | 50<br>100 | 15<br>30 | B-B<br>B-B |

Figure 1-17 Twin-fitted hose assemblies (note: A-size for small torches, B-size for medium and heavy duty torches). (*Courtesy of Linde Co., Division of Union Carbide*.)

abuse, and to be unaffected by either oxygen or acetylene gas.

A good hose is made in four layers: a high quality gum rubber inner liner, then two layers of strong fabric, and finally, a tough rubber outer covering. The oxygen and acetylene hoses are either fastened together by vulcanizing or by clips to avoid tangling (see Figure 1-17). Most acetylene hoses are colored red with left-handed grooved ferrules or nuts. The oxygen hoses are either green or black and have right-handed ferrules or nuts (see Figures 1-18, 1-19).

NOTE: Do not attempt to use an acetylene hose for oxygen because a combustible mixture could result; do not switch the fittings.

*CAUTION: Avoid dropping sharp objects or hot metal on the hoses. A rupture in the hose could cause a fire. Avoid kinking the hoses because it will hinder the flow of gas.*

Be careful not to set the cylinder cart tray down on the hoses. The sharp edges of the tray can cut a hose or restrict the flow of gas. Hoses are made in several sizes, inside diameter (ID)

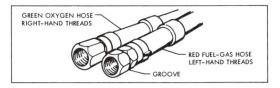

Figure 1-18 Welding hoses and fittings. (*Courtesy of Linde Co., Division of Union Carbide*.)

3/16", 1/4", and 5/16" (4.76 mm, 6.35 mm, and 7.93 mm) (see Figure 1-17). The hoses are made in various lengths; the most commonly used lengths are 12½ ft. and 25 ft. (3.17 m and 6.35 m). Never repair a broken line with a piece of copper tubing because the tubing could cause a chemical reaction.

## SAFETY CHECK VALVES

A device that must be included with all oxygen and acetylene equipment is a back-fire device called a Safety Check Valve. This device permits gas to flow to the torch but does not allow gas to flow back from the torch—which is called backfires. This device will reduce the

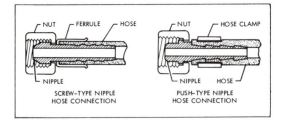

**Figure 1-19** Welding hoses and fittings, cross section view. (*Courtesy of Linde Co., Division of Union Carbide.*)

danger to the torch, hoses, and regulators. It is either attached near the torch or near the regulators (see Figure 1-20).

## CONSTRUCTION AND OPERATION OF THE TORCH

The torch, sometimes called a blowpipe, is a precision engineered piece of equipment. It is sometimes considered the most important part of the welding equipment because it directs the flame. In the torch, the oxygen and acetylene gases are mixed in the proper proportions. The mixture is burned at the end of the tip allowing the operator to direct the flame to the work. The torch is divided into four parts (see Figure 1-21):

Body—which serves as a handle

Valves—which control the amount of gases desired

Mixing head or chamber—where the gases are mixed in proper proportions

Tips or tubes—combustion takes place at the end of the tip.

The torches are made of several different types of material such as brass and aluminum. Size and capacities vary from small torches for welding sheet metal or soldering to large ones used where great amounts of heat are necessary or for cutting heavy steel. Most body shops use either small or medium sized torches. The

smaller the torch, the easier it is to handle in restricted places. The medium sized torches are generally used in repair or maintenance shops.

There are two types of torches—injector and equal pressure. In the injector type, the acetylene at low pressure is carried through the torch by the higher pressure oxygen flowing from a jet or venturi (see Figure 1-21). In the equal pressure type the gases pass through the torch at equal pressure; the pressure is great enough to force the gas through the mixing chamber. All torches are equipped with a pair of needle valves or ball type valves that serve two purposes. One is to turn the gases on or shut them off, while the other is used to make the proper adjustment to obtain the desired flame (see Figure 1-22).

NOTE: Care must be taken not to close the valves too tightly when shutting off the gas. This could result in damage to the valve needle and seat.

Sometimes an operator may experience difficulty in holding the proper flame adjustment. If faulty gauges are not the cause and if the connections are tight, a loose packing seal nut is the problem. If the valve moves very freely, the

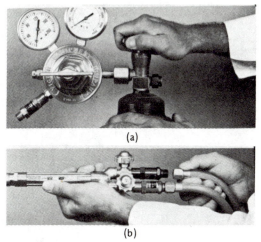

(a)

(b)

**Figure 1-20** Safety check or reverse-flow check valves can be installed on the regulator (a) or at the torch (b). (*Courtesy of Victor Welding and Cutting.*)

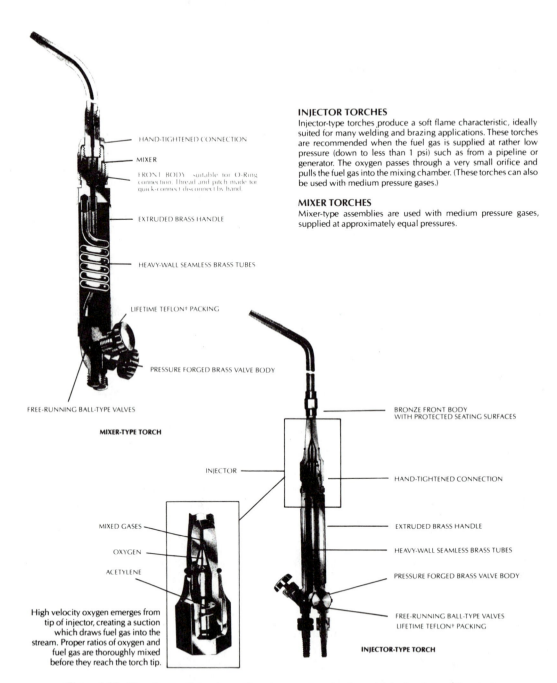

HAND-TIGHTENED CONNECTION

MIXER

FRONT BODY suitable for O-Ring connection. Thread and pitch made for quick-connect disconnect by hand.

EXTRUDED BRASS HANDLE

HEAVY-WALL SEAMLESS BRASS TUBES

LIFETIME TEFLON† PACKING

PRESSURE FORGED BRASS VALVE BODY

FREE-RUNNING BALL-TYPE VALVES

**MIXER-TYPE TORCH**

## INJECTOR TORCHES

Injector-type torches produce a soft flame characteristic, ideally suited for many welding and brazing applications. These torches are recommended when the fuel gas is supplied at rather low pressure (down to less than 1 psi) such as from a pipeline or generator. The oxygen passes through a very small orifice and pulls the fuel gas into the mixing chamber. (These torches can also be used with medium pressure gases.)

## MIXER TORCHES

Mixer-type assemblies are used with medium pressure gases, supplied at approximately equal pressures.

BRONZE FRONT BODY
WITH PROTECTED SEATING SURFACES

INJECTOR

HAND-TIGHTENED CONNECTION

MIXED GASES

OXYGEN

ACETYLENE

EXTRUDED BRASS HANDLE

HEAVY-WALL SEAMLESS BRASS TUBES

PRESSURE FORGED BRASS VALVE BODY

High velocity oxygen emerges from tip of injector, creating a suction which draws fuel gas into the stream. Proper ratios of oxygen and fuel gas are thoroughly mixed before they reach the torch tip.

FREE-RUNNING BALL-TYPE VALVES
LIFETIME TEFLON† PACKING

**INJECTOR-TYPE TORCH**

**Figure 1-21**  Two types of torches: mixer type or equal mix and injector type. (*Courtesy of Linde Co., Division of Union Carbide.*)

14

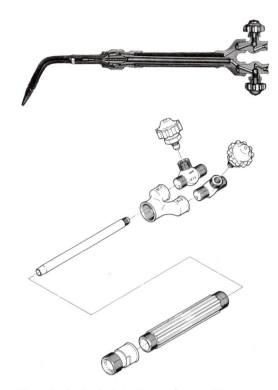

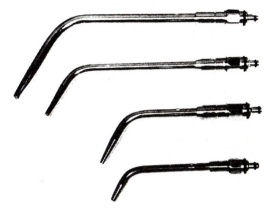

**Figure 1-23** Welding and heating tip or tubes with mixing chambers. (*Courtesy of Linde Co., Division of Union Carbide.*)

**Figure 1-22** Exploded view of a welding torch. (*Courtesy of Victor Welding and Cutting.*)

packing seal nut under the valve knob should be slowly tightened until there is a slight drag or stiffness. Occasionally, the seal nuts are too tight and it is difficult to turn the torch on. This can be remedied by backing off slightly on the packing seal nut.

## MIXING CHAMBER

Some torches have the mixing chamber in the torch only. Many torches have the mixing chamber combined with the tip and head assembly. Each time the tip is changed, the mixing chamber is changed also. This permits the mixing chamber to supply a sufficient amount of mixed gases for a certain size tip. Each mixing chamber is drilled with holes or orifices to match the size tip to be used. Some are stamped on the sides showing what size

tips can be used—either size numbers or letters. These holes or orifices must not be altered or enlarged because they are precision drilled to permit the proper amount of gases to pass through. Figure 1-21 is a cutaway view of mixing chambers for both types of torches.

## TIPS OR TUBES

The tips or tubes are either permanently soldered or removable from the mixing head or chamber. All tips are made of copper (see Figure 1-23). Tips are made in various sizes and numbered or lettered according to the manufacturer. These tips are calibrated to permit a certain amount of mixing gases to pass through at a given pressure for a given length of time. Normally, the lower the number, the smaller the tip, but some manufacturers use the tip drill size as a numbering device. For example, a number 2 tip would be the same as a number 62 in another make of torch. See Table 1-1 for comparison. (Also see Figure 1-24.)

These tips average from 5 to 7 lbs. (34 to 48 kPa) for acetylene (gauge reading). For oxygen, 5 to 7 lbs. (34 to 48 kPa) gauge reading. For 1/4" (6.350 mm) metal, up to 12 lbs. (82 kPa) for oxygen. Consult the manufacturer's instruc-

**Table 1-1**

**Comparison Chart**

| MANUFACTURER OF TORCH | 20 ga (1.2m) | 1/16" (1.580mm) | 1/8" (3.175mm) | 3/16" (4.763mm) | 1/4" (6.350mm) |
|---|---|---|---|---|---|
| Airco | 0 | 1 | 3 | 4 | 5 |
| Smith | B60 | B61 | B63 | B64 | B65 |
| Purox | 4 | 6 | 12 | 15 | 20 |
| Linde | 2 | 3,4 | 5 | 7 | 8 |

tions for the proper pressure to be used for certain tips.

Tip cleaners (Figure 1-24b) are used to remove carbon and other matter from the tip.

## WELDING GOGGLES

The welder must wear suitable welding goggles when welding, brazing, and cutting with

| | For Welding Metal Thickness | | †Head Size |
|---|---|---|---|
| | in. | mm | |
| | 1/32 to 1/16 | 0.8 to 1.6 | 4 |
| | 1/16 to 3/32 | 1.6 to 2.4 | 6 |
| | 3/32 to 1/8 | 2.4 to 3.2 | 9 |
| W-17 | 1/8 to 1/4 | 3.2 to 6 | 15 |
| (chrome | 1/4 to 1/2 | 6 to 13 | 30 |
| plated) | 1/2 to 3/4 | 13 to 19 | 55 |
| | 3/4 to 1 | 19 to 25 | 70 |
| | | | 100 |
| | Heating | | 150 |
| | | | 250 |
| | up to 32 ga | 0.2 | 1 |
| | 32 ga to 25 ga | 0.2 to 0.5 | 2 |
| | 25 ga to 1/32 | 0.5 to 0.8 | 4 |
| | 1/32 to 1/16 | 0.8 to 1.6 | 6 |
| | 1/16 to 3/32 | 1.6 to 2.4 | 9 |
| W-201 | 1/8 to 3/16 | 3.2 to 4.8 | 15 |
| (un- | 1/4 to 3/8 | 6 to 9.5 | 30 |
| plated) | 1/2 to 5/8 | 13 to 16 | 55 |
| | 3/4 to 1 | 19 to 25 | 70 |
| | Over 1 | 25 | 100 |
| | Or Heating | | |
| | up to 32 ga | 0.2 | 1 |
| | 32 to 25 ga | 0.2 to 0.5 | 2 |
| | 1/32 | 0.8 | 4 |
| W-200 | 1/32 to 1/16 | 0.8 to 1.6 | 6 |
| (un- | 3/32 to 1/8 | 2.4 to 3.2 | 9 |
| plated) | 1/8 to 3/16 | 4.8 to 6 | 15 |
| | 3/8 | 10 | 30 |

(a)

a torch. The welding flame produces light and heat rays that can damage the eye tissue; continued exposure will result in the loss of sight. Sparks and molten metal that fly from the weld or cutting area can cause painful eye injury. Brazing, although performed at a lower temperature, gives off harmful rays from the flame and the molten flux. These protective devices are available in goggle type, eye shields with 2 X 4 lenses that can be used with eye glasses and the full face shield (see Figure 1-25). The full face shield is recommended for cutting and welding overhead to protect the face from flying sparks and metal. For most welding and cutting, lenses with a shade number from 4 to 6 are recommended. For welding and cutting heavy metal, shades numbered from 6 to 8 are recommended. A clear plastic lens should be inserted in front of the welding lens to prevent the welding lens from being damaged by flying sparks and molten metal. If the welding lens be-

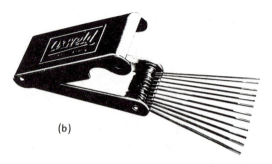

(b)

**Figure 1-24** (a) Tip selection guide for different sizes of torches (head size number indicates approximate flow of acetylene in cfh at normal operating pressures); (b) Used for cleaning the tip of carbon and other matter. (*Courtesy of Linde Co., Division of Union Carbide.*)

SINGLE FLINT LIGHTER

THREE FLINT LIGHTER

**Figure 1-26** Always use a striker or lighter to light the torch. Do not use matches or cigarette lighters to avoid burnt fingers. (*Courtesy of Linde Co., Division of Union Carbide.*)

(a)

|  | LINDE Part No. |
|---|---|
| **No. 24 Coverall Goggle** | |
| with Shade 3 lenses | 713F28 |
| with Shade 4 lenses | 713F21 |
| with Shade 5 lenses | 713F27 |
| with Shade 6 lenses | 713F22 |
| with Shade 8 lenses | 713F23 |
| **No. 800H Coverall Goggle,** large size | |
| with Shade 5 lenses | 737F79 |
| **No. 850H Coverall Goggles,** large size, | |
| with flip up lens holder and Shade 5 lenses | 737F77 |
| **No. 42 Flexible Plate Goggle,** stationary front, | |
| with Shade 5 lenses | 639104 |
| **No. 43 Flexible Plate Goggle,** lift front, | |
| with Shade 5 lenses | 639105 |

† To order other lenses or plates see pages 11 and 12

(b)

**Figure 1-25** (a) Different styles of welding goggles; (b) welding goggles and lenses. (*Courtesy of Linde Co., Division of Union Carbide.*)

comes severely chipped, the amount of protection from the lens will decrease.

Many welders wear eye glasses and the welding glasses hinder the line of sight, especially if the eye glasses are bi-focals. An equipment manufacturer has produced magnifiers that fit into the eye shield type goggles. They are available in various strengths. Consult local distributors for additional information.

Another important piece of welding equipment is a striker (see Figure 1-26). The striker or flint lighter is a device used to ignite the gases flowing from the welding torch. The striker is the simplest and safest means of lighting a torch.

## OPERATION OF THE TORCH

### *SETTING UP THE TORCH*

In the previous sections of this welding chapter, the various parts of the oxy-acetylene torch were discussed. Now we will discuss how to connect them together. Securely fasten the oxygen and acetylene cylinders in a welding cart or fasten to a sturdy wall, vertical column, etc. (see Figure 1-27) with chains, straps or other mechanical means. Remove caps from cylinders and crack open cylinder valves to clean out any possible dirt (see Figure 1-28). To crack open means to momentarily open and close the cylinder valve slightly. Check cylinder outlet for damaged threads or chipped outlet seat.

Figure 1-27 Securely fasten cylinders in a cart or to a sturdy wall. (*Courtesy of Victor Welding and Cutting*.)

Figure 1-28 Crack open cylinder valves to clean out any possible dirt. Wipe connection fitting with a clean cloth; check for damaged threads. (*Courtesy of Victor Welding and Cutting.*)

NOTE: Use the wrench provided with the torch or use the correct size wrenches. Do not use pliers, vise grips, or pipe wrenches because they damage the soft metal fittings.

After inspecting the regulator coupling, attach the proper regulator and tighten (see Figure 1-29). With the regulator screw turned out, or loosened, slowly open the cylinders and check for leaks (Figure 1-30). Use pure ivory soap suds for checking leaks. Attach the safety check valve on the regulator, if available. Inspect hoses and coupling for damage. Attach the proper hose and coupling on the regulator (red on acetylene, left-handed threads; green on oxygen, right-handed threads) and tighten (see Figure 1-31). Inspect the torch connection, then attach the carrier or welding hose to the torch and tighten (see Figure 1-32). Most torches are marked O, OX, or Oxygen on the oxygen inlet fitting and A, AC, or Acetylene on the acetylene inlet fitting. With the valve knobs facing upward, the acetylene is generally on the left side, oxygen on the right. Install the proper tip if it is not already on the torch (see Figure 1-33). With the torch valve closed, slow-

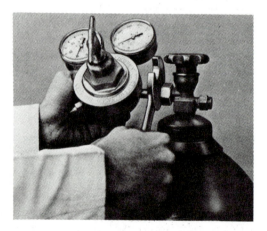

Figure 1-29 Attach regulator using the proper wrench. (*Courtesy of Victor Welding and Cutting.*)

Figure 1-30 With regulator screw backed out, stand to one side and slowly open cylinder valve. Check for leaks with approved solution. All the way open for oxygen, 1/4 to 1/2 turn for acetylene. (*Courtesy of Victor Welding and Cutting.*)

gauges decreases, this indicates a leak. With no leaks and the proper tip installed, the torch is ready for use.

## *HOW TO LIGHT A TORCH*

This is a suggested method of lighting a torch. Open the oxygen cylinder all the way, SLOWLY. Do not stand facing the gauge, stand to one side, in case the diaphragm ruptures. Turn the regulator screw in so the gauge reads about 5 psi (34 kPa). Open the torch valve to fill the line and shut off the torch valve. Slowly open the acetylene valve one-fourth to one-half turn. In case of fire, it is easier to shut off the cylinder. Turn the

Figure 1-31 Attach green or black hose to the oxygen regulator (right-handed threads) and the red hose to the acetylene regulator (left-handed threads). (*Courtesy of Victor Welding and Cutting.*)

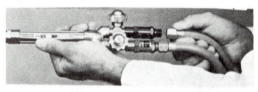

Figure 1-32 Attach the green hose to the oxygen inlet of the torch and the red hose to the acetylene inlet or fitting. Most torches are marked O, OX, or Oxygen, and A, AC, or Acetylene on the inlet fittings. (*Courtesy of Victor Welding and Cutting.*)

(a)

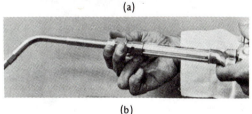

(b)

Figure 1-33 (a) Select proper welding tip assembly, check O-ring before installing; (b) tighten hand-tight—do not use wrenches on most torches. (*Courtesy of Victor Welding and Cutting.*)

ly open the cylinder valve, turn in the regulator screws so the low pressure gauges read from 5 to 7 psi (34 to 48 kPa), and check for leaks with soap suds. Another way to check for leaks is to shut off the cylinder valve with the torch valves closed. If the pressure indicator on the regulator screw in so that the gauge reads about 5 psi (34 kPa). Open the torch valve to fill the lines and then shut off the torch valve. Put on the welding goggles. Crack open the oxygen valve (Figure 1-34) and open the acetylene valve about one-half turn (Figure 1-35). Light the torch with a flint lighter or striker.

NOTE: Do not use matches or cigarette lighters because the resultant flare-up could burn your fingers.

Open the acetylene more until the flame starts to leave the welding tip (Figure 1-36), then back off slightly to bring the flame back to the tip (Figure 1-37). Open the oxygen more until the desired flame is reached, as in Figure 1-38.

NOTE: Some welders will differ with the above procedure. For example, they may crack the oxygen valve before opening the acetylene

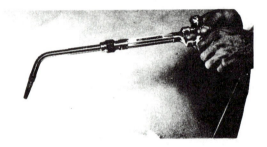

Figure 1-36 Open acetylene more until the flame starts to leave the welding tip. (*Courtesy of Victor Welding and Cutting.*)

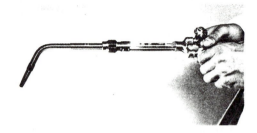

Figure 1-37 Back off slightly to bring the flame back to the tip. (*Courtesy of Victor Welding and Cutting.*)

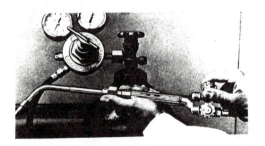

Figure 1-34 Crack open oxygen torch valve a fraction of a turn. (*Courtesy of Victor Welding and Cutting.*)

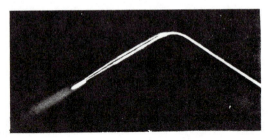

Figure 1-38 Open oxygen valve until the desired flame is reached. As the bright inner cone appears and the feathery flax disappears, that is a neutral flame. (*Courtesy of Victor Welding and Cutting.*)

Figure 1-35 Open acetylene torch valve 1/2 turn, light torch with flint lighter or striker. Do not use matches or a cigarette lighter. (*Courtesy of Victor Welding and Cutting.*)

torch valve. This will eliminate the carbon particles in the air when they are lighting the torch.

To extinguish the torch turn off the acetylene torch valve first, then the oxygen valve. The purpose of shutting off the acetylene first is to avoid black smoke; in addition, the oxygen will clean the torch from any carbon residue. After finishing welding, the following procedures should be followed. Close the acetylene cylinder valve. Open the acetylene torch valve and drain lines. Shut off the torch valve and unscrew the regulator adjusting screw (counterclockwise) to relieve pressure from the regulator spring and diaphragm. Close the oxygen cylinder valve. Open the oxygen torch valve and drain line. Shut off the torch valve and unscrew the regulator adjusting screw (counterclockwise) to relieve pressure from the regulator spring and diaphragm. Hang the hose on the cart or wall brackets. Keep hoses off the floor, away from wheels, and from underneath the cylinder cart tray.

## SELECTION OF WELDING TIPS

The selection of welding tips is very important. For example, in body and fender work, most of the welding is done on about 20-gauge sheet metal about 1/16" (1.5 mm). Care must be exercised in welding to avoid burning holes in the metal. Since the orifice size, or hole, in the tip determines the amount of acetylene and oxygen fed to the flame, the orifice therefore determines the amount of heat produced by the torch. If the tip orifice is too small, not enough heat will be produced for proper flow of the molten metal. If the orifice is too large, too much heat will be produced and will result in poor welds with excessive penetration. Occasionally, a certain size tip or a particular welding job heats the metal to the melting point, but the metal seems difficult to melt and the puddle is difficult to maintain. This indicates that the torch tip must be larger in order to heat the metal and also provide additional heat to supply the necessary "latent heat" or "heat of fusion."

LATENT HEAT: the amount of additional heat required to change a given object from a solid to liquid after the melting point is

reached. One indication that a tip is too small for a given job is the length of time needed to establish a molten spot or puddle. If it is too long a period (more than 10-15 seconds), the tip is too small.

Each manufacturer of torches will recommend a different pressure for a similar size tip. For most purposes in welding sheet metal, that is 20 gauge (about 1.5 mm), 5 to 7 psi (34 kPa) of acetylene and 5 to 7 psi (34 to 48 kPa) of oxygen will generally do the job. A #0 tip from one torch may be about the same as a #2 from another brand of torch. The heavier the metal, the more intense heat required. Experienced welders will have few problems in selecting tips. If you are not sure, consult the torch manufacturers for the correct tip and pressure. See Table 1-1 for tip selection.

## THE OXY-ACETYLENE FLAME

The oxy-acetylene flame does the work of the oxy-acetylene equipment. The flame has been sometimes called the "master of metals." The various parts of the welding equipment are set up to serve the flame by supplying the proper amounts of gases to it. The operator of the torch has full control as to efficiency, size, and quality of the flame. The welder must learn to adjust the flame properly to do a good job; a poorly adjusted flame could even result in destroying the metal's characteristics.

## TYPES OF FLAMES

There are three types of flames, depending on the ratio of acetylene and oxygen (see Figure 1-39).

**Neutral Flame** This flame consists of approximately equal parts of acetylene and oxygen combined in the inner cone to produce a flame of about 5700°F (3154°C). It is called neutral because it does not have any chemical effect on the molten steel; it does not add or take anything from it. The inner cone of the flame is a light blue. This is surrounded by an outer envelope of the flame, which is generally medium or darker blue. This outer envelope protects the molten steel from contamination of oxygen and other gases in the atmosphere. The neutral flame is used most of the time when welding mild steel.

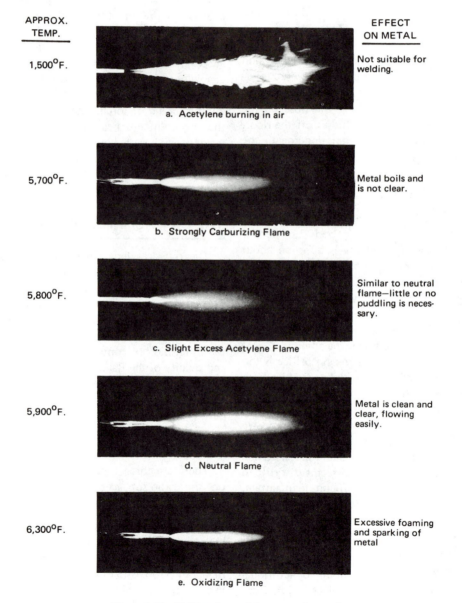

| APPROX. TEMP. | | EFFECT ON METAL |
|---|---|---|
| 1,500°F. | a. Acetylene burning in air | Not suitable for welding. |
| 5,700°F. | b. Strongly Carburizing Flame | Metal boils and is not clear. |
| 5,800°F. | c. Slight Excess Acetylene Flame | Similar to neutral flame—little or no puddling is necessary. |
| 5,900°F. | d. Neutral Flame | Metal is clean and clear, flowing easily. |
| 6,300°F. | e. Oxidizing Flame | Excessive foaming and sparking of metal |

Figure 1-39 Welding flame. (*Courtesy of Airco.*)

**Carburizing or Reducing Flame** This flame has an excessive amount of acetylene. It will cause a feather of pale blue extending from the inner cone. Welding with a carburizing flame causes the excess carbon to combine with the molten metal, producing a very hard and brittle weld. As more acetylene is added to the flame, the outer envelope becomes longer. One use of a carburizing flame with a long envelope is for applying body solder. It is sometimes termed a soft flame because of a great reduction in heat and a less concentrated flame. Some authorities recommend the carburizing flame for brazing and hard soldering whereas many welders disagree and use a neutral flame for brazing. Some job conditions require a definite size or length of the acetylene feather in relation to the inner cone. A 2X flame means that the feather is twice as long as the inner cone.

**Oxidizing Flame** This flame is produced by excessive oxygen in the flame. The inner cone is smaller and sharper. The outer envelope is shorter and produces a hissing sound. The flame is undesirable for welding because it has a tendency to oxidize the steel. This produces a noticeable foaming of the weld, and the weld is very weak. This flame can reach as high as 6300°F (3482°C) and with high heat, the oxygen combines readily with other metals.

## CHEMISTRY OF THE OXY-ACETYLENE FLAME

In the oxy-acetylene neutral flame, the combustion of the acetylene and oxygen form carbon monoxide. The carbon monoxide combines with the oxygen in the atmosphere to form carbon dioxide and water vapor. It may be noted that when welding is done in a very confined area, the torch may go out. This is caused by lack of oxygen in the area which was used up by the welding (and the welder).

## QUESTIONS

1. How safe is the oxy-acetylene torch?

2. What were some early uses of the acetylene flame?

3. What is the highest temperature of the oxy-acetylene flame?

4. How did the early small shops produce their own acetylene?

5. How is acetylene gas produced?

6. How is calcium carbide made?

7. What is the highest pressure at which acetylene can be stored in a free state or in an open container?

8. Why is acetylene dangerous beyond that certain pressure?

9. Is leaking acetylene easily detected?

10. How is commercial oxygen obtained?

11. What type metal is used in the manufacture of oxygen cylinders?

12. What is the normal pressure of a full oxygen cylinder?

13. What is the purpose of the protective cap on the cylinder?

14. What type valve is used on the oxygen cylinder?

15. What safety device is used to protect the cylinder in case of excessive pressure caused by heat?

16. What are the two most commonly used cylinder sizes?

17. Why is oil or grease dangerous around any part of the welding set?

18. What are the two things found in an acetylene cylinder?

19. What is the purpose of acetone in the cylinder?

20. What safety devices are used for protecting the acetylene cylinder in case of high heat?

21. What type threads are used on the acetylene valve outlet?

22. How is acetylene gas sold?

23. How should be acetylene cylinder be stored?

24. Name another gas that can be used for welding, cutting, and brazing?

25. Is MAPP gas safer than acetylene gas?

26. What is used to reduce the high cylinder pressure down to the correct welding pressure?

27. What are the two gauges on the regulator used for?

28. What will cause a regulator to increase in pressure after the torch is shut off?

29. What is the difference between a single and two-stage regulator?

30. What type outlet fittings are found on the oxygen and acetylene regulators?

31. How are the torch hoses attached together to prevent tangling?

32. How can the acetylene fitting be identified?

33. What is the purpose of the torch or blow pipe?

34. Name the four parts of the torch.

35. What are the two types of torches?

36. Does the mixing chamber usually identify the tips to be used?

37. What are welding tips made of?

38. Do all of the torch manufacturers use the same tip number?

39. Can the light rays from welding damage the eyes?

40. Why are welding goggles a must when using the cutting torch?

41. What shade numbers are recommended for welding, cutting, and brazing?

42. What should be done to the cylinder before attaching the regulator?

43. Can a pair of pliers or a pipe wrench be used for tightening the regulators?

44. What can be used for checking leaks?

45. What is another way to check for leaks?

46. Should matches or cigarette lighters be used to light the torch?

47. Why should the acetylene torch valve be shut off first rather than the oxygen?

48. What determines the size tip to be used?

49. What is latent heat?

50. What are the three types of flames?

51. Which type flame does not have any chemical effect on the molten metal?

52. Which type flame causes the weld to foam?

# OXY-ACETYLENE STEEL WELDING

## chapter 2

## WELDING PROCEDURES—STEEL

The oxygen-acetylene steel welding process is a method of joining or fusing pieces of metal together by heating the metal to the melting point so that it will flow or intermingle together. After a person has become familiar with the welding equipment and with the practice of lighting the torch and adjusting the flame, the next step is welding. A good weld depends on several factors including the proper selection of a tip, proper torch adjustment, correct movement of the torch, and correct speed.

The most important thing in becoming a good welder and making strong welds is

PRACTICE. Too many beginners feel that practice is boring after a short time. It takes time to master the proper technique of the torch movement, making puddles, running beads with filler rods, etc.

Most welding books or articles generally picture welding on flat surfaces and on clean metal. In most fields, a welder often cannot hold a torch at a 40° to 60° angle and cannot follow many other recommended procedures. A good repair person is expected to weld in any position, in confined places, and on metal that is rusty or painted. This is a why beginning welder should do most of the practice welding on old panels or scrap and use clean metal only in the beginning. A lot of the welding performed by a repairman is on metal similar to autobody sheet metal, which is approximately 20-gauge metal, less than 1/16" thick. This is very thin metal and is more difficult to weld than heavier metal because the welder can easily burn a hole through it and heat can warp it.

When starting to practice, the first step is to make puddles—commonly called puddling. A puddle is a small spot of molten metal. You will need pieces of sheet metal about 4" X 6" (101.6 X 152.4 mm) of 20 gauge, No. 2 Purox, No. 1 Marquette, or equivalent size. The size of practice metal is only a suggestion. Large pieces may be wasted. If heavier metal is used, then the tip size must correspond with the thickness of the metal. Refer to Table 1-1.

Light the torch and adjust it to a neutral flame. Acetylene and oxygen pressure will vary according to the type of torch used. Hold the torch tip so that the inner cone is approximately 1/8" (3.175 mm) from the metal. Hold the tip at a 45° to 60° angle in line with the direction of travel at the beginning. Hold the torch flame in one spot until a pool of molten metal forms [about 3/16" to 1/4" (4 .76 mm to 6.35 mm) in diameter on light metal], then move the puddle forward (see Figure 2-1). This should be done at an even speed in order to keep the puddle at uniform width. A circular or weaving back and forth motion can be used. One of the objects in puddling is to move forward in a straight line. Observe the penetration of the puddle. If it is too deep (when the base metal starts to melt through), several things can be done to correct this. Increase speed, or readjust the flame by decreasing oxygen and acetylene, or decrease the angle of the torch. If penetration is too little, decrease the speed, or adjust the torch for more heat, or increase the angle of the torch. Make four to five puddles the length of the metal. Observe the results, especially width and penetration. Look at the underside of the metal to check for correct penetration: it should be slightly raised under the puddle. If metal shows high ridges, the metal is melting through. Use several pieces of metal until you have complete control of the torch with even, straight puddles.

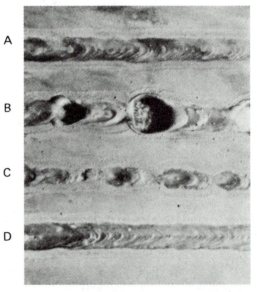

A

B

C

D

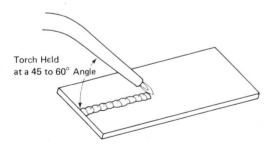

Torch Held at a 45 to 60° Angle

Figure 2-1 Running a bead or puddle without the use of filler rod. Note: The above illustration depicts a left-handed welder going from left to right. A right-handed welder would go from right to left. (*Courtesy of Linde Division of Union Carbide*.)

Figure 2-2 Sample of good and bad beads or puddles without the use of filler rod. A and D are good beads or puddles; B. Overheating, resulting in holes in the metal; C. Insufficient heat and poor forward motion. (*Courtesy of Linde Division of Union Carbide*.)

One indication of the correct speed would be a puddle about 1/4" (6.35 mm) wide with even, half-round ripples. If the ripples are pointed, generally it means too much speed. Too slow speed will cause extra wide puddles and excessive penetration (see Figure 2-2 for good and bad beads). Vary the torch angle and note the results. A good bead shows an even line of circles or ripples over the complete puddle. A bad bead shows high and low spots and an uneven line of travel.

## FUSION WELDS

Another form of puddling is to make a butt joint, which is joining two pieces of metal placed end to end. Using the same technique as in puddling, with a round or weaving motion, melt both edges of the metal so that they flow together (see Figure 2-3). Care must be taken not to burn the edges through. Another exercise is to place two pieces of metal parallel with each other so that one or more edges are even with the other (see Figures 2-4 and 2-5). Fuse the edges together. Proper torch technique is important so that you do not melt the edges too much. The fusion weld should be even. Once again, the key to mastery of the torch is practice.

## WELDS MADE USING FILLER ROD

After practicing running puddles and gaining torch control, the next step is to add filler rod to the puddle to form a bead. Filler, or welding rod is used to provide additional strength to metal joints where the weld is. Where strength is needed or when two pieces of metal do not fit closely together, a welding rod is used. This is generally used in places where fusion welds cannot be performed. Filler rods are made in various sizes from 1/16" (1.58 mm) in diameter for light metal to 1/4" (6.35 mm) in diameter for heavy plate steel and are generally made in 36" (914.4 mm) lengths. For general purpose welding, the American Welding Society #RG45

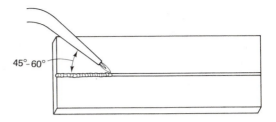

**Figure 2-3** Fusing a butt joint without the use of filler rod.

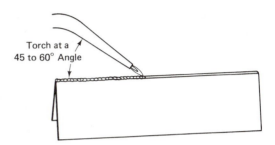

**Figure 2-4** Fusing an outside corner joint without the use of filler rod.

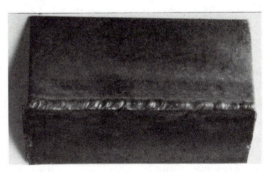

**Figure 2-5** Fusing an outside corner joint without filler rod.

rod is used. These are copper coated, produce excellent welds, and are good for general purpose welding. The AWS #RG60 bare rod can also be used when higher strength is needed. It is self-fluxing with a sufficient amount of silicon and magnesium to remove impurities from the molten metal. Always bend one end of the welding rod, to avoid picking up the hot end, and to avoid poking someone's eye out.

Selection of the proper size rod is very important in obtaining a satisfactory weld. A filler rod that is too small will cause difficulty in controlling the puddle or it will cause the problem of having to add enough filler rod to the puddle to make a good weld. On the other hand, if the rod is too large, it will cool the puddle down and the weld bead will lack proper penetration, which will make it higher than necessary. A good weld or bead should have good fusion, good penetration, and a good consistent bead with a slightly convex contour (if welding on a flat surface). With the use of a filler rod, proper coordination or manipulation between the torch and the filler rod is necessary. Once again, practice is the secret to mastering welding.

**Procedure** Use 20-gauge (1 mm) metal pieces approximately 4″ X 8″ (101.6 mm X 203.2 mm), a small tip (depending on the torch) and 1/16″ (1.588 mm) steel welding rod. The method using filler rod is very similar to the puddling technique. This is a general method. Light the torch and adjust it to a neutral flame. Heat a spot to form a puddle, then place the rod into the puddle, but do not melt the rod with the torch flame. Add just enough rod to form a slightly convex or raised bead. Move the puddle forward adding rod at a consistent rate (see Figures 2-6 and 2-7). If the weld becomes too high, it means too much rod has been added. Also, the excess rod may have cooled the puddle too much. If the bead is too low, more rod should be added.

NOTE: Some beginners may have a tendency to melt the rod with the torch rather than in the puddle. This generally results in poor penetration and poor looking welds.

Run several beads and check the height of the beads and the consistency of the width of the bead. As mentioned before, when making puddles, the ripples in the bead will give an indication of speed. Correct speed will result in

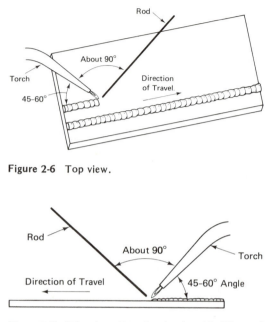

**Figure 2-6**   Top view.

**Figure 2-7**   Side view. Running beads with filler rod and the recommended torch and rod angle.

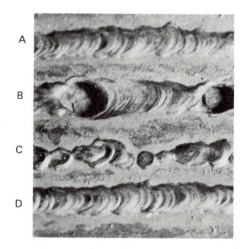

**Figure 2-8**   A and D welds show the proper procedure with fairly uniform width, height and ripples; B. Too much heat was used, causing holes and very wide bead formation, uneven speed of travel, mostly too slow; C. Not enough heat was used to melt the base metal properly; the filler is not fused sufficiently with the base metal; uneven travel. (*Courtesy of Linde Division of Union Carbide.*)

even, half-round ripples with slightly convex beads and good penetration (see Figure 2-8a and d). If the speed is too fast, the ripples will be pointed with the bead higher than normal and there will be poor penetration. If speed is too slow, the bead will be lower than normal with too much penetration and width (see Figure 2-8b).

## WELDING JOINTS

Most of the oxy-acetylene welding performed involves some type of joint. Some of the joints involved are called butt, lap, inside corner (or fillet), or outside corner, sometimes called edge welds (see Figure 2-9). If the repairman is involved in repairing auto bodies or truck cabs, all of the panels join by one type of joint or another. The beginning welder should spend most time practicing the type of joints he will probably find on the job. Also, various thicknesses of metal should be used in order to overcome the problems of joining heavy and light metals together.

Some authorities recommend using only clean metal. In the repair situation, the metal will not always be clean. For the beginner's first welding practices, the joint should be made on clean metal; then he should practice with metal that is most often welded on the job.

One of the very common types of joints welded is the butt joint. Because the metal is very thin, distortion from heat will be a common problem. For practice welding, the conditions should be ideal with clean metal, proper gap, etc. When doing an actual job involving, for example, splicing two pieces of metal together, the gaps between the two pieces may be unequal and not ideal. This is where the previous practice and torch control will play a large part in the success of the welding operation. Also, the proper torch angle will be hard to maintain. When making a butt joint, a

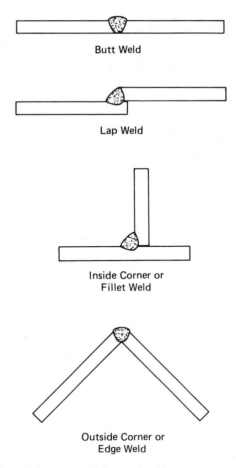

Butt Weld

Lap Weld

Inside Corner or
Fillet Weld

Outside Corner or
Edge Weld

**Figure 2-9** Types of joints and welds.

gap of at least 1/16" (1.58 mm) should be present before welding to avoid warping from the expansion of the metal.

## PROCEDURE

Practice on two pieces of 20-gauge (1 mm) metal approximately 2" X 6" (50 mm X 203 mm); use oxy-acetylene with #1 or #2 tip depending on manufacture and 1/16" (1.58 mm) welding rod. If heavier metal is used, the tip should be larger. Position the pieces of metal together using either of two methods. One method uses an equal gap of about 1/16"

(1.58 mm) (see Figure 2-10); the other method uses a tapered gap (see Figure 2-11) with one end closed and the other end about 1/8" (3.17 mm) apart. When using equal gap, tack weld both ends, then tack weld about every 2" (50.8 mm). Then proceed to weld the joint. Observe the type and penetration of the bead. When using the tapered method, start from the closed end. Use vise grips to hold the pieces of metal in position when welding. In both cases, hold the torch at approximately a 45° to 60° angle with the tip pointing in the direction of travel.

The torch should melt the edges to be joined of the two pieces of metal evenly, and the rod should be added at a uniform rate. After completing the weld, observe the bead and penetration (see Figure 2-12a). For further checking, place one side of the metal in a vise with the weld about 1/4" (6.35 mm) from the top of the jaw. Bend the metal back and forth and if the weld cracks, the weld is poor (see Figure 2-12b). If it breaks along the side of the weld, the weld is good. When welding light gauge metal, it may be advisable to weld about 1" (25.4 mm), skip 2" (50.8 mm), and then weld 1" (25.4 mm). Another method is to

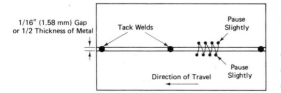

Figure 2-10  Butt weld, tack weld method—the gap cannot close due to tack welds.

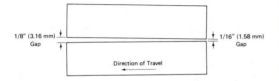

Figure 2-11  Butt weld, tapered method—the tapered gap allows for expansion. The gap will close as the weld progresses across the joint.

**Figure 2-12(a)**  Example of a butt weld. (*Photo by S. Suydam.*)

quench the metal with cool water when the metal has changed from red to black. This will prevent the heat from expanding the surrounding panel. If the panel is miscut and there is an extra wide gap, several beads may be necessary to fill the gap.

## LAP JOINTS

The lap joint, where one piece of metal overlaps another, is the most common joint used in the manufacture of cars. In some respects, it is easier to weld a lap joint than other joints but other problems will arise. The problem of expansion is less because the one panel can expand over the other. The lap joint is perhaps more difficult to weld because of the difference in the surface area of the two pieces of metal to be welded. The edge of the top piece requires less heat than the bottom one. Beginners have a tendency to apply heat to the bottom piece to form the proper puddle. The results are a good-looking bead on the top piece with no fusion to the bottom one.

Welding lap joints involving two different thicknesses of metal requires a different pro-

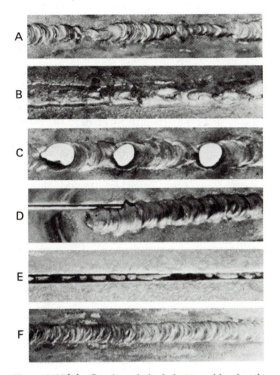

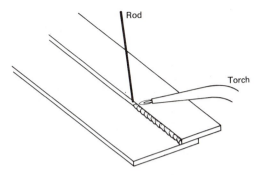

**Figure 2-13** Lap weld with the rod between the torch and the edge of the top piece.

**Figure 2-12(b)** Good and bad butt welds. In the above welds, "A" shows "skips" which are the result of irregular timing or too rapid torch advance. "B" shows excessive penetration, caused by too slow torch advance. "C" shows holes burned through the result of too much heat—either flame too close or advance too slow. "D" shows an off-center weld which is simply the result of not pointing the flame along the seam. "E" is the result of insufficient melting and lack of penetration on reverse side. "F" demonstrates the way a proper weld should look. The width of the weld is uniform. The ripples are evenly spaced. (*Courtesy of Linde Co., Division of Union Carbide.*)

cedure. The torch should be aimed directly at the heavier gauge metal—the lighter metal will also heat up to the puddle stage. If the flame is directed toward the joint, the light gauge metal will melt away first while the heavier piece will be too cool to puddle.

In many cases involving lap joints on panels, the upper or lower pieces may separate due to expansion from the heat. Some repairmen will use draw rivets or sheetmetal screws or clamps to hold the metal together before welding.

However, the problem here is if force is used to hold the two edges together, the panel can easily be distorted. To help eliminate distortion, welds of 1" (25.4 mm) and skip 2" (50.8 mm) plus cooling of the weld with water, should be used. Placing wet asbestos or cloths along the edge of the joint will keep the heat from traveling out into the surrounding areas of the panel. Once the panel is welded the full length, go back and weld in between the first welds. When welding a panel or light gauge metal, it is not advisable to weld a continuous bead the full length of the joint because severe warping could result.

Use two strips of 20-gauge (1 mm) metal, 2" X 4" (50.8 X 101.6 mm) or several pieces cut out of scrap metal panel (the metal can be heavier). Use an oxy-acetylene torch set with #1 or #2 tip, larger if heavier metal is used; a 1/16" (1.58 mm) steel welding rod and vise grips. Position the metal to form a lap joint. Use vise grips to hold the metal in place. Start to weld by concentrating most of the heat on the bottom piece until the proper puddle is formed. Then add the filler rod and move the torch slightly toward the top edge (see Figures 2-13 and 2-14). The puddle and filler metal will follow toward the top edge, and the top edge will puddle and fuse to the lower piece. Add the filler at a uniform rate and be careful

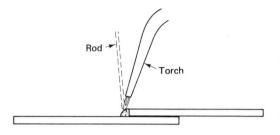

**Figure 2-14** End view of a lap weld.

not to melt the top piece too fast. If the top edge melts away too fast, remove the torch for a second and then add filler rod to the melted edge filling in the hole. Then, proceed along the joint. After completing the weld, observe the bead formation, the penetration, melting of the top edge, and the uniformity of the weld (see Figure 2-15). Place the lower section in a vise and using vise grips, bend the top piece back and forth until it breaks. Observe where the break is. If it broke in the joint, the weld was improper. If the metal broke next to the weld, the weld is strong.

Try this procedure again until this type of joint is mastered. An example of another problem is welding two different size exhaust pipes together with a rather large gap between the two pipes. The filler rod must be used to bridge the gap and have the proper fusion to both pieces.

## T-WELD OR FILLET WELD

Another type of weld joint frequently performed is the T-joint weld, or fillet weld. This weld is sometimes called the inside corner weld. Some of the precautions mentioned in lap welding are the same for the T-joint. The bottom plate will require more heat for forming a puddle than the vertical one will. If too much heat is just applied to the vertical piece, it will melt away before the bottom is hot enough. It is very important to establish a puddle on the lower piece before adding the filler rod or else there will be little or no fusion. Add the filler rod close to the vertical piece, and tip the torch slightly toward the vertical piece. The puddle will form on the vertical piece, and the filler rod will fuse to it (see Figure 2-16). As the bead advances, be careful not to undercut the vertical piece. This can be avoided by adding the filler rod to the bead nearest to the vertical piece. A slight weaving motion of the torch from the bottom to the top piece and back again with a slight pause on the bottom will aid in the proper formation of the bead. If you melt a hole in the vertical piece, immediately fill it with filler rod and move the torch back and forth to keep the hole from getting any larger.

On many occasions, the vertical piece is heavier than the flat or horizontal one.

**Figure 2-15** Example of a lap weld. (*Photo by S. Suydam.*)

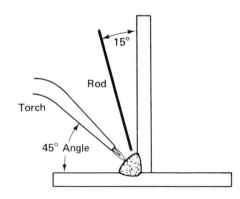

**Figure 2-16** Torch and rod position for inside corner or fillet welds.

Here, the technique will vary where more heat is required to form the puddle on the vertical one rather than on the flat piece. Tip the torch toward the vertical piece to avoid melting a hole in the flat piece. With practice and experience, these different situations that arise will cause fewer problems as time goes on.

Repairing a bicycle frame or welding a piece of pipe vertically to a flat piece of metal will present a little variation in technique because the bead will be circular rather than a straight one. The proper torch angle should be maintained with the constant change around the circle.

NOTE: Welding in corners may result in improper torch flame action, caused by the lack of sufficient oxygen in the corner, sometimes called spitting. By increasing the oxygen slightly, this will sometimes eliminate the problem but the oxygen must be reset again after welding in the corner is finished or else oxidation of the future welds could cause problems.

Take two pieces of 20-gauge (1 mm) metal or body metal scrap, vise grips, oxyacetylene torch outfit, 1/16″ (1.58 mm) steel welding rod, and fire brick or some other means to support the vertical piece of metal. Position the metal to form a T-joint and tack weld both ends. Start from the right if right-handed or the left if left-handed. Be sure to establish the proper puddle on the bottom or flat piece, then add the filler rod. Using a slight weaving motion, move toward the vertical piece and establish the proper fusion between the filler rod and the vertical piece. Continue the length of the joint, being careful not to undercut the vertical piece. Observe the weld after completion for contour of the bead (should be slightly concave); check penetration and even distribution of the filler rod on both pieces (see Figure 2-17). Place the vertical piece in a vise and bend the bottom piece down on the side of the bead or weld using vise grips.

**Figure 2-17** Sample of an inside corner or fillet weld. (Photo by S. Suydam.)

Bend it back and forth until it breaks and observe the location of the break. If it broke through the weld, it was improperly welded.

## WELDING LIGHT GAUGE METAL IN VERTICAL, HORIZONTAL, AND OVERHEAD

During the previous discussion about using butt, fillet (T-joint), and lap weld joints, when making repairs, the position of the welded area was frequently mentioned. The law of gravity will play an important part when welding out of position or doing other than flat welding. Although many of the areas to be welded either on vehicles or for general repairs are unfortunately either in a vertical, horizontal, or overhead position, the basic joints and technique are the same. If a welder has mastered the proper technique in the previously discussed joints, welding out of position will not create serious problems. However, there are a few points that should be remembered.

Gravity will have a tendency to force the molten steel to run downward. When welding out of position, keep the molten puddle to a smaller size than when welding on a flat position. However, the proper penetration and rod will help to control the size of the puddle.

A lower heat setting on the torch or one size smaller tip may be necessary to control the size of the puddle yet maintain the proper penetration.

## VERTICAL

In vertical welding, welding downward could cause a problem with the molten metal trying to run ahead of the puddle. Also vertical down bead will be generally smaller in size (see Figure 2-18a). Vertical welding up beads are generally a little wider in size, and sometimes become too large because of the molten metal trying to run back over the bead (see Figure 2-18b). Both up and down vertical welds will become easier with practice.

## HORIZONTAL

When welding horizontal, the molten metal will have a tendency to run down overlapping the lower edge of the joint but will lack the proper penetration and fusion; this is often

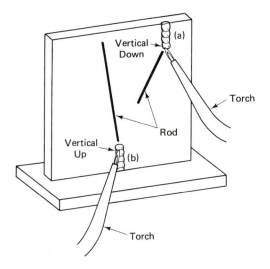

Figure 2-18 (a) Torch and rod position for vertical down welding; (b) torch and rod position for vertical up welding.

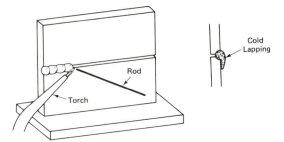

Figure 2-19 Torch and rod position for horizontal welding.

called cold lapping. Keeping a smaller puddle and pointing the torch tip a little bit upward will also help to keep the metal in place (about 5° below the horizontal plane (see Figure 2-19). Also, using the next size heavier welding rod and adding the rod to the upper edge of the puddle will help to keep the molten metal in place.

## OVERHEAD

Overhead welding requires a smaller puddle that is still large enough to get the desired penetration and fusion (see Figure 2-20). This can be done by reducing the size of the welding flame or using the next size smaller welding tip. Using a little larger welding rod will help to keep the puddle to the proper size. The force of the welding flame will also help to keep the molten metal in place. Many welders are afraid of welding overhead, but with a little care, it is a very safe position.

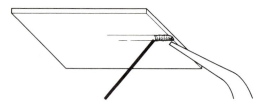

Figure 2-20 Torch and rod position for overhead welding.

CAUTION: *Never position yourself direct-ly beneath the welding area if the situation demands that the welder lay on a creeper or directly underneath, while making repairs.*

Move off to one side and avoid having the arm underneath the area. Remove as much loose dirt, rust, or other foreign materials as possible. Sparks and molten metal will fall downward and could cause serious burns. Bend the welding rod in the middle at right angles to bring the hand holding the rod out of the way of falling metal and sparks. Wear gloves, and a full face shield, rather than goggles. Check the area for flammable materials to avoid fires.

## QUESTIONS

1. What are some of the things necessary to make a good weld?
2. Is it possible to always hold the correct torch angle?
3. What is a puddle in welding terms?
4. How close should the inner blue cone be held to the metal?
5. How can the lack of penetration be corrected?
6. What is one indication of corrected torch speed?
7. What is fusion without filler rod?
8. What are the two types of steel welding rods generally used?
9. Why is it necessary to bend one end of the welding rod?
10. What happens if too much filler metal is added to the puddle at one time?
11. What is the indication of welding too rapidly?
12. What type metal should be used for practice?
13. Why is it necessary to space the metal apart when butt welding?
14. What is the tapered method of butt welding?
15. When lap welding two pieces of metal of equal thickness, which requires the most heat?
16. How can the two pieces of metal be held together when welding panels?
17. When welding inside corners, where should the filler rod be added to prevent undercutting?
18. Why is welding inside corner welds on pipe or tubes different from welding on flat pieces?
19. What must be taken into consideration when welding in a vertical or over-head position?
20. What is "cold lapping?"

# HEAVY PLATE

## chapter 3

### STEEL WELDING—PLATE STEEL

Welding plate steel with an oxy-acetylene welder requires a little more preparation but is welded the same way as light gauge metal. Plate steel is generally referred to as sheet steel; it has a thickness from 1/8" to 3/16" (3.175 mm to 4.76 mm). For any large amount of welding on plate steel, it is economical to use an electric arc or a MIG welder. However, when an arc welder is not available, the oxy-acetylene torch can be used, but it takes more time.

The biggest problem with plate steel is to get the proper penetration so that the joint will be strong enough. Another problem often en-countered when plate welding with a torch is that the torch tips are too small to supply sufficient heat. Therefore, the welder increases the gauge pressures and adjusts the torch to the maximum heat output. When this occurs, the puddle and filler rod are blown out of the way and it is difficult to form a good bead. This is very common when welding in a vertical or horizontal position. Also, the chance of under-cutting the plate is very great (see Figure 3-1). During this kind of welding, the torch is generally pointed in the direction of travel, but in heavy plate welding and in some cast iron welding, it is helpful to use a backhand motion (see Figure 3-2). Backhand motion is where the tip and flame are directed back over the welded

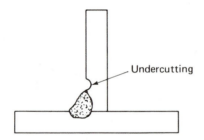

Figure 3-1 Undercutting.

bead. This motion tends to anneal the welded area and relieve stresses that may develop in the weld area. It also helps the welder to form a better bead and improve penetration.

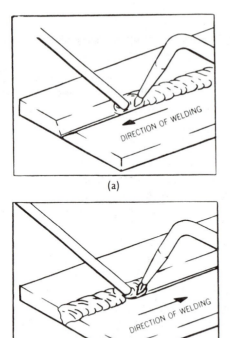

(a)

(b)

Figure 3-2 (a) Forehand welding and (b) backhand welding. (*Courtesy of Victor Welding and Cutting.*)

(a)                    (b)

Figure 3-3 (a) Bevel or V-joint and (b) bevel or V-joint with land.

When preparing plates for welding, the edges should be beveled at a 45° angle to insure adequate penetration. This is especially true for butt welding, which is commonly called a V-joint (see Figure 3-3). If the plates are over 3/8" (9.52 mm) thick, it is better to put a double 45° bevel on the ends of the plate. "Double bevel" is beveling the edge on both sides, commonly referred to as a double V-joint (see Figure 3-4).

Beveling can be done by grinding with a disc grinder or using a cutting torch. The torch can be guided with a piece of angle iron to get the correct angle. Refer to the oxy-acetylene cutting section for cutting bevels (also see Figures 4-21, 4-22, and 4-23 in Chapter 4). On all butt welds, the two pieces must be spaced apart before welding to eliminate distortion from expansion during the welding process. A common rule is to space the plate apart one half the thickness of the plate to be welded, unless the edges are beveled.

## PROCEDURE

To practice, you will need an oxy-acetylene torch, 3/32" and 1/8" (2.81 mm and 3.17 mm) welding rods, steel plate 2" X 4" (50.8 mm X 101.6 mm), and a grinder or cutting torch for beveling. Use plate greater than 3/16" (4.76 mm) thick. Grind or torch cut two edges of two plates to be joined. Align the two plates together leaving a 1/16" (1.58 mm) gap between them. Select a welding tip suitable for the thickness of the metal. Adjust the torch to a neutral flame.

Tack weld the two plates at both ends. Start to weld from one end establishing a

Figure 3-4   Double bevel or double V-joint.

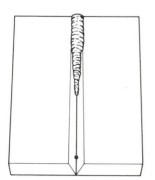

Figure 3-5   Heavy plate with multi-bead welding.

## LAP JOINTS ON HEAVY PLATE

Welding lap joints on heavy plates is very similar to lap welds on light gauge metal. The biggest problem is to get sufficient penetration on the lower plate without burning away too much of the edge on the top plate. The bottom plate will require about two-thirds more heat for proper fusion than the top plate. The resulting bead should be convex and the sides of the bead even.

You will need an oxy-acetylene welding torch with suitable tip, 3/32" or 1/8" (2.31 mm or 3.17 mm) steel welding rod, steel plate 2" X 4" (50.8 mm X 101.6 mm), and 3/16" (4.76 mm) thick steel. Position the plate with a 1" overlap. Adjust the torch to a neutral flame and direct the torch to the bottom plate to form a puddle next to the edge of the plate. As the bottom plate is being brought up to the puddle stage, the upper piece will also become hot. As soon as the puddle is formed, add the filler rod, and using a circular or weaving motion of the torch, move the torch back and forth across the lap joint. Continue to add the filler rod to both pieces as the torch moves back and forth. Pause slightly on the lower plate because it requires more heat to maintain the puddle. Continue across the length of the joint and observe the shape of the bead and the amount of the penetration. Observe the finished bead and the contour, and the uniform shape of the bead. Repeat the procedure but use a backhand motion and notice the difference between the two in the shape of the bead.

puddle on both pieces by using a circular or a weaving motion covering the full width of the bevel. Then add the filler rod to the puddles being sure that the filler rod is fusing to both pieces. Continue across the joint and avoid adding too much filler rod to the joint to avoid having high convex beads. The bead should be slightly convex and the sides of the bead even. Repeat this procedure using a backhand motion and notice the difference in the bead.

On heavy or thick plates, it may be necessary to make several beads to fill the joint (see Figure 3-5).

## QUESTIONS

1.  What happens if a welding tip is set to the maximum output when welding steel plate?

2.  What is back hand welding?

3.  Why is it necessary to bevel heavy plates?
4.  What is a double V joint?
5.  What is a basic rule for spacing plates?

# OXY-ACETYLENE CUTTING

## chapter 4

The oxy-acetylene cutting torch is very useful for cutting off rusted bolts, severing heavy steel plates, removing mangled sheet metal, and many other useful applications. Heavy iron castings over 1 ft. (304.8 mm) thick can be cut with a cutting torch. Stainless steel and some special types of steel cannot be cut.

The cutting torch is one of the fastest ways to cut ferrous metal. However, there are three main objections: the width of the cut is too great, the edges of the metal must be cleaned off in order to make a good weld, and there is danger of fire. The cutting torch is very widely used throughout industry, either by manual operation or automatic machine cutting using many torches at once. The oxy-acetylene cutting is actually a burning process. It is the severing of ferrous metal by heat and oxygen. After the ferrous metal is heated to the kindling temperature, the addition of oxygen will cause the metal to burn at a very rapid rate. This process is called oxidation of the ferrous metal—the same process that destroys many thousands of automobiles and other objects made of ferrous metal. The most common term is rusting, but rusting is oxidation at a much slower rate. For example, heat a piece of ferrous metal and then expose it to a moist atmosphere. Immediately, the metal will start to oxidize. The cutting process is a chemical process involving the chemical affinity of oxygen for ferrous metals; the affinity increases when the metal is heated. The hotter the metal, the greater the affinity.

## THE CUTTING TORCH

Most of the torches used for cutting are welding torches equipped with a cutting head in place of the welding tip (see Figure 4-1). Also, many of the cutting torches (especially the larger ones) are combination torches. There is a difference due to the absence of one of the two oxygen valves. One valve takes care of volume and the final adjustment of the flame (see Figures 4-2 and 4-3). All cutting torches have a high pressure lever that is used to feed the high pressure oxygen to the cut area.

The tip of the cutting torch has a series of orifices around the outside of the tip. The average cutting tip has either 4, 6, or 8 orifices called preheat jets (see Figure 4-4). The size of the jets vary for different thicknesses of metal to be cut. The center jet, which is generally larger than the outside ones, is the cutting or oxygen jet. The tips also come in different

**Figure 4-2** (a) High pressure lever. Heavy duty cutting torch with one oxygen valve. Cuts up to 12" (305 mm) thickness of steel. (*Courtesy of Linde Division of Union Carbide.*)

**Figure 4-3** Heavy duty cutting torch with extra long reach. Can cut up to 30" (762 mm) of steel. Used often for the demolition of buildings or cutting I-beams. (*Courtesy of Linde Division of Union Carbide.*)

shapes, angles and jet arrangement for special cutting, such as scaling, rivet cutting or gouging (see Figure 4-5). A tip used for cutting light steel plate or sheet metal will have one preheat jet and an oxygen jet (see Figure 4-5a).

**Selection of the Proper Cutting Tip** As stated before, most of the common cutting tips have either 4 or 6 preheat jets. When using a 4-jet tip, 2-jets and the center jet must be in line with the cut (see Figure 4-6). With a 6-jet tip, the arrangement of the jet in relation to the line of the cut is not important.

For beveling steel plate, see Figure 4-7 for preheat jet alignment.

## *PROPER CUTTING PRESSURES*

The amount of acetylene and oxygen pressure to be used is determined by the size of the tip to be used, which is determined by the thickness of the metal to be cut. Most tips found on the average cutting torch will cut up to about 2" to 3" (50.8 mm to 76.2 mm) of steel. The small torches, called aircraft torches, will cut up to about 3/4" (19.05 mm) of steel. The average cutting job will be under 1/2" (12.7 mm).

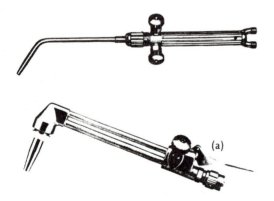

**Figure 4-1** (a) High pressure lever. The cutting assembly fits on the torch body after the welding tip is removed. This light duty torch and cutting assembly cuts up to 2" (50.8 mm) of steel. (*Courtesy of Linde Division of Union Carbide.*)

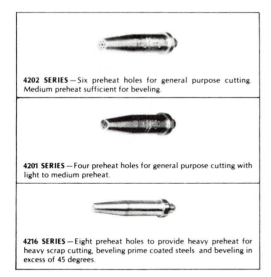

**4202 SERIES** — Six preheat holes for general purpose cutting. Medium preheat sufficient for beveling.

**4201 SERIES** — Four preheat holes for general purpose cutting with light to medium preheat.

**4216 SERIES** — Eight preheat holes to provide heavy preheat for heavy scrap cutting, beveling prime coated steels and beveling in excess of 45 degrees.

| Metal Thickness | | SERIES | | | | | |
|---|---|---|---|---|---|---|---|
| | | 4202 | | 4201 | | 4216 | |
| In. | mm | Size | Part No. | Size | Part No. | Size | Part No. |
| 1/8 | 3 | 3 | 16K08 | 2 | 638869 | 3 | 998589 |
| 1/4 | 6 | | | 3 | 16K05 | | |
| 1/2 | 13 | 4 | 16K09 | 4 | 16K06 | 4 | 998590 |
| 3/4 | 19 | 5 | 16K10 | 5 | 16K07 | | |
| 1 | 25 | | | | | | |
| 1-1/2 | 38 | | | | | 6 | 998591 |
| 2 | 50 | | | | | | |
| 2-1/2 | 64 | 7 | 16K11 | | | 8 | 998592 |
| 3 | 75 | | | | | | |
| 4 | 100 | | | | | | |
| 6 | 150 | 9 | 16K12 | | | 10 | 998593 |
| 8 | 200 | 11 | 16K13 | | | | |
| 10 | 250 | | | | | 12 | 998594 |
| 12 | 300 | 13 | 16K14 | | | | |

**Figure 4-4** Cutting tip selection guide. (*Courtesy of Linde Division of Union Carbide.*)

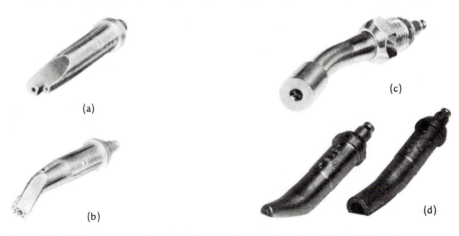

**Figure 4-5** Different types of cutting tips: (a) sheetmetal cutting; (b) grooving and gouging; (c) rivet cutting; and (d) cutting and trimming, especially used for close to a restricting surface. (*Courtesy of Linde Division of Union Carbide.*)

The pressure setting for oxygen and acetylene should follow the torch manufacturer's recommendations, if known. But for the average torch, 20 psi (138 kPa) of oxygen and 5 psi (34 kPa) of acetylene can be used for the average cutting. Cutting of metal beyond 1" (25.4 mm) thick may require additional oxygen. Many medium-sized torches will cut up to 6" (152.4 mm), and the oxygen may have to be raised to 100 psi (690 kPa).

NOTE: When cutting heavy pieces of metal, over 1" (25.4 mm) thick, the metal must be cut all the way as it is very difficult to start the cut again, if stopped in the middle.

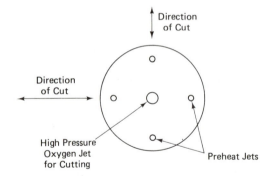

**Figure 4-6** Alignment of preheat jets for cutting.

The approximate 4 to 1 ratio [20 psi (138 kPa) of oxygen to 5 psi (34 kPa) of acetylene] produces a fine cutting or burning action. In lighting the torch, 5 psi (34 kPa) of acetylene and 5 psi (34 kPa) of oxygen are used to produce a netural flame for preheat of the metal. The remaining 15 psi (103 kPa) of oxygen is needed for the burning, or oxidation, of the metal with enough velocity of the oxygen to wash or blow out the oxidized metal from the cut area. One of the most common mistakes made by inexperienced torch operators is that they do not set the oxygen pressure high enough. This generally produces a poor cutting

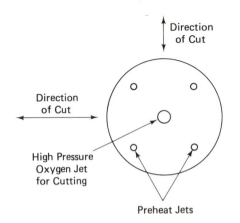

**Figure 4-7** Alignment of preheat jets for beveling.

action; most of the time the molten metal flows back in behind the tip, re-fusing the two pieces together again. When cutting thicker pieces, there is difficulty cutting through the metal. There is also a problem of using more oxygen than needed, which will produce too much oxidized metal (slag) especially at the bottom of the cut. Also, the cut or kerf becomes too wide, and the cut is uneven.

*KERF:* a cut or slot from which the metal has been removed by the cutting torch in a flame cutting operation.

## SAFETY IN CUTTING

Using a cutting torch presents two major problems: fire and molten metal. Before using the cutting torch, check the surrounding area for any (in)flammable material such as thinner or gas soaked rags, paper, open containers of flammable liquids, car batteries, and anything that will burn. When cutting outdoors, be careful of dried grass or leaves.

*CAUTION: Remember that the sparks and molten metal will travel as far as 30 ft. (about 10 meters) from the torch (see Figure 4-8).*

The person using the torch will be preoccupied with cutting and not notice anything burning nearby (see Figure 4-9). Cutting off body panels can present a problem because of the large amount of flammable material in an automobile, such as upholstery, body sealer, and wiring. Remove as much of this material as possible and cover the rest with wet rags. Undercoating and sealer can be scraped off. Be careful around machinery especially with areas of accumulated grease or fuel oil.

The torch operator must wear welding goggles or face shields (see Figure 4-10). Sneakers or fabric shoes are not recommended. The molten metal being blown from the cut will burn through. Pants without cuffs are recommended. Shirt pockets should be buttoned and all bookmatches or propane lighters removed.

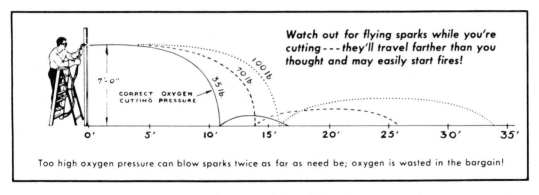

*Watch out for flying sparks while you're cutting---they'll travel farther than you thought and may easily start fires!*

Too high oxygen pressure can blow sparks twice as far as need be; oxygen is wasted in the bargain!

**Figure 4-8** Sparks. (*Courtesy of Meco Modern Engineering Co.*)

Gloves should be worn because the hot metal and undercoating or other material will drip from around the cut area. Keep welding hoses clear of falling metal and burning materials. Double check for any smoldering material before you leave the area. Many fires have started after the worker has left for home. Mark HOT on any recent cut metal to avoid having other persons burn their fingers. Keep welding tanks away from the line of sparks and molten metal.

*CAUTION: Safety precautions are not foolish notions. Keep a fire extinguisher and water handy.*

When attempting to cut apart a container, be sure you know what type of liquid was stored in it. Many different types of flammable liquids leave a residue after the liquid has been removed. The heat of the cutting torch will cause it to give off an explosive vapor. There are certain methods to make the container safe to cut, but that should be left to the professionals. Cutting of galvanized metal such as old water tanks can be dangerous. When it is

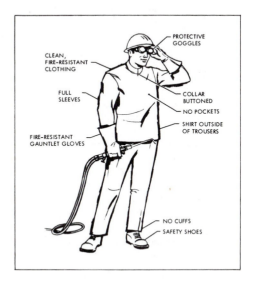

**Figure 4-9** Prepare surrounding area for cutting. (*Courtesy of Linde Division of Union Carbide.*)

**Figure 4-10** Select clothing to provide maximum protection from sparks and hot metal. (*Courtesy of Linde Division of Union Carbide.*)

necessary to cut them, make sure it is in a very well ventilated area preferably outside; if possible, have the wind toward your back (see Figure 4-11).

NOTE: Do not cut directly on a concrete floor because the heat will cause small pieces of concrete to pop off the surface. Also, avoid having large amounts of slag fall on the floor at one time. Have a pail of water or sand directly under the cutting area to catch the molten slag, or if possible, thoroughly wet the floor before cutting.

*CAUTION: Inhaling fumes can cause very serious respiratory illness.*

## PROCEDURES

Install the cutting assembly on torch handle, if required. Select the proper cutting tip and line up the jets if you are using a 4-preheat jet tip. [Most cutting assemblies have one tip supplied with the purchase, but other tips are available from the supplier. Medium-sized cutting torches are equipped with a tip that will cut up to 1½" (38.1 mm) while the light duty torches are equipped with a tip for up to 3/4" (19.05 mm) thickness cuts.]

Open up the oxygen tank all the way and set the regulator at 20 psi (138 kPa). Open the acetylene tank valve 1/4 turn and set the regulator at 5 psi (34 kPa). Crack open both torch valves to purge the lines; recheck the regulator settings. Check the surrounding area for any potential fire hazards and remove them. Don't forget the fire extinguisher and water buckets.

Open the acetylene valve about one-half turn and light the torch with a striker. Continue to open the acetylene valve until the flame starts to leave the tip; then close the valve

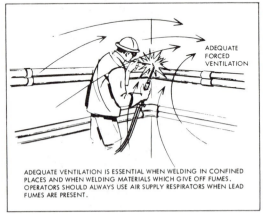

ADEQUATE VENTILATION IS ESSENTIAL WHEN WELDING IN CONFINED PLACES AND WHEN WELDING MATERIALS WHICH GIVE OFF FUMES. OPERATORS SHOULD ALWAYS USE AIR SUPPLY RESPIRATORS WHEN LEAD FUMES ARE PRESENT.

**Figure 4-11** Proper ventilation is important for both welding and cutting. (*Courtesy of Linde Division of Union Carbide.*)

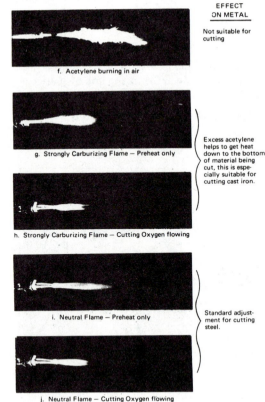

EFFECT ON METAL

f. Acetylene burning in air — Not suitable for cutting

g. Strongly Carburizing Flame — Preheat only

h. Strongly Carburizing Flame — Cutting Oxygen flowing

Excess acetylene helps to get heat down to the bottom of material being cut, this is especially suitable for cutting cast iron.

i. Neutral Flame — Preheat only

j. Neutral Flame — Cutting Oxygen flowing

Standard adjustment for cutting steel.

**Figure 4-12** Cutting flame. (*Courtesy of Airco.*)

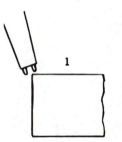

1. Start to preheat; point tip at angle on edge of plate.

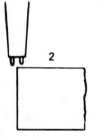

2. Rotate tip to upright position.

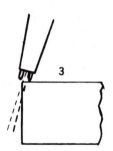

3. Press oxy valve slowly; as cut starts, rotate tip backward slightly.

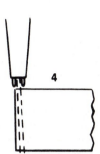

4. Now rotate to upright position without moving tip forward.

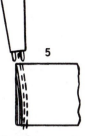

5. Rotate tip more to point slightly in direction of cut.

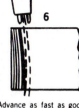

6. Advance as fast as good cutting action will permit.

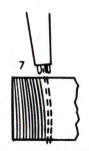

7. Do not jerk; maintain slight leading angle toward direction of cut.

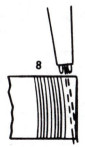

8. Slow down; let cutting stream sever corner edge at bottom.

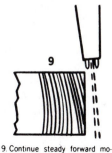

9. Continue steady forward motion until tip has cleared end.

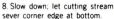
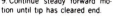

**Figure 4-13** Recommended procedure for efficient flame cutting of steel plate. (*Courtesy of Victor Welding and Cutting.*)

slightly until the flame returns to the tip. Open the oxygen torch valve all the way. Next, start to open the cutting assembly oxygen valve until a neutral flame appears. Next, depress the cutting lever slightly and notice the flame. If the flame is still neutral, the torch is ready for cutting (see Figure 4-12). If a carbon feather appears, add additional oxygen until the neutral flame appears. If a combination welding and cutting torch is being used, the oxygen valve is used for adjusting the flame. Shut off by closing the acetylene torch valve first. This will put out the flame and oxygen will cleanse the tip of any carbon. Next, close the oxygen torch valve and the valve on the cutting assembly. Practice lighting the torch again and adjusting the flame.

Mark a line with soapstone about 1" (25.8 mm) from one end of a piece of steel plate. Center punch the line if the plate is rusty. Place the plate so that the line is off the edge of the bench or table. Put the welding goggles on and relight the torch and adjust to a neutral flame. Hold the torch vertical to the edge of the plate above the soapstone line with the blue flame cones about 1/8" (3.17 mm) from the metal (see Figure 4-13). Preheat the metal to a bright red, then slowly depress the oxygen lever until the metal starts to burn away (see Figure 4-14). Move slowly forward making sure that the metal burns and is blown out of the cut or kerf (see Figure 4-15). Tilt the tip of the torch forward about 5° to 10° and continue across the plate to the other end. The cut piece should drop onto the floor. If the piece appears to be cut but does not drop off, tap it with a hammer as some slag remained in the cut.

*CAUTION: Do not use the torch as a hammer.*

Only use as much oxygen as is necessary to cut the metal. There is no need to depress the lever all the way down if 1/4 of the way is doing the job. Observe the cutting marks on the edge of the plate. They should be a fine, even

**Figure 4-15** Move cutting tip forward at an even rate of speed for a clean cut. (*Courtesy of Victor Welding and Cutting.*)

line. The cut itself should be straight, and the side even. If the side appears ragged, this generally indicates that the speed was too slow. Mark another line and cut the plate again to develop the proper speed and to cut in a straight line. Also, use different thicknesses of plate to further improve your cutting technique. For really straight cuts, clamp a piece of angle iron on the plate and use it for a torch guide (see Figure 4-16).

See Figures 4-17 and 4-18 for cutting problems and their solutions.

Many times the cut must be started away from the edge of the plate—like cutting out round or odd-shaped pieces or cutting holes for bolts. It is more difficult to start the cut. It is necessary to hold the torch in one spot a little longer until the metal starts to turn red (see No. 1, Figure 4-19). Then raise the torch tip to

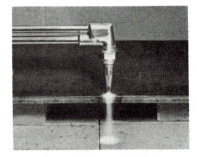

**Figure 4-14** Heat metal to a bright red and slowly depress the oxygen lever to start the cut. (*Courtesy of Victor Welding and Cutting.*)

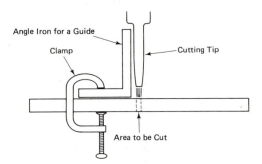

**Figure 4-16** Fixture used for straight cutting.

| | Top Edge | Bottom Edge | Plate Face Condition and Drag Line Pattern | Sound of Cut | Slag Pattern | Possible Drop Cut |
|---|---|---|---|---|---|---|
| Quality cut – all adjustments correct – see Fig. 1 | Clean and square with no roll over. No slag on top of plate. | Square and free of slag. | Surface is smooth and clean. Drag lines show uniform, vertical pattern. Plate requires no additional processing. | Smooth and regular – | Regular – Consistently vertical through length of cut. | Yes |
| Travel speed too fast – see Fig. 2 | Relatively clean and square. | Considerable slag adheres to bottom edge – cutting oxygen stream moving too fast to allow complete oxidation. | Occasional gouges appear, drag lines are pronounced, and slant away from direction of cut. | No noticeable sputtering. | Irregular pattern as cutting oxygen stream intermittently lags behind the position of tip. | No |
| Travel speed too slow – see Fig. 3 | Rough and uneven. Slightly melted away due to excess pre-heat exposure. | Considerable slag adheres to bottom edge. | Upper portion is clean and smooth. Lower section severely gouged due to wandering oxygen stream. | Erratic – noticeable sputtering. | Irregular and erratic due to uneven progress of oxygen stream | 50/50 |
| Cutting oxygen pressure too high – see Fig. 4 | Uneven; out of square. Excessive amount of top edge oxidized as oxygen stream expands upon entry. | Relatively clean and square – free of slag. | Plate is relatively free of pits or gouges but draglines irregular and erratic due to excessive oxygen stream turbulence. | Smooth and regular but exceptionally loud. | Distinct and regular due to force of oxygen stream. | Yes |
| Cutting oxygen pressure too low – see Fig. 5 | Generally clean and square. | Considerable slag adheres to plate as cutting oxygen has difficulty penetrating metal. | Plate is fairly smooth but much slag adheres to bottom. Draglines slant away from cut. | Irregular and occasional sputtering. | Irregular and weak – oxygen pressure not sufficient to carry through metal. | No |
| Too much pre-heat – see Fig. 6 | Rounded edge produced by excessive heat. Molten beads also deposited on top of plate. | Moderate amount of slag. Usually adheres to bottom edge. | Draglines are fairly regular and smooth. Excessive metal being removed from plate leaves much slag on bottom edge. | Regular and even but louder than normal (higher flow on pre-heat). | Regular and consistent. | Yes |
| Too little pre-heat – see Fig. 7 | Top edge slightly rounded and out of square. | Often irregular – moderate amount of slag may appear. | Pits and gouges sometimes appear. Draglines uniform and well defined. | Erratic and uneven. | Erratic and uneven. | Not Normally |
| Tip too far from plate – see Fig. 8 | Flared and partially blown away – out of square. | Relatively even – little if any slag adhesion. | Smooth and even. Draglines are uniform and vertical. | Smooth and even – constant. | Smooth and even. | Yes |
| Tip too close to plate – see Fig. 9 | Generally rough due to pre-heats interrupting cutting oxygen stream. | Relatively even and slag free. | Occasional gouges will result from pre-heat popping. Draglines show irregular pattern. | Relatively even sputtering. | Usually regular. | 50/50 |

NOTE: Cutting tip may also be damaged if allowed to contact plate.

Figure 4-17  Oxy-fuel cutting reference chart. Part I—description of variables/conditions. (*Courtesy of Victor Welding and Cutting.*)

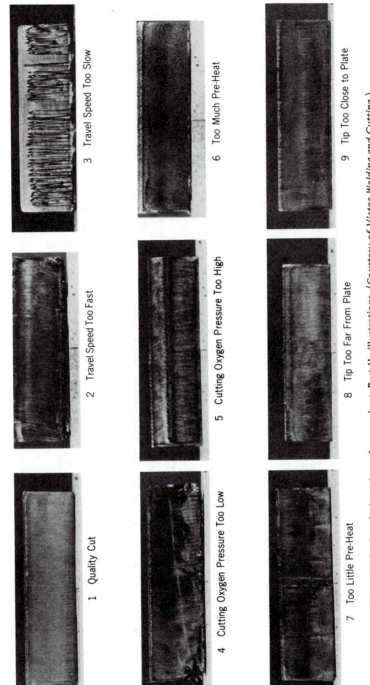

**Figure 4-18** Oxy-fuel cutting reference chart. Part II—illustrations. (*Courtesy of Victor Welding and Cutting.*)

1 Quality Cut

2 Travel Speed Too Fast

3 Travel Speed Too Slow

4 Cutting Oxygen Pressure Too Low

5 Cutting Oxygen Pressure Too High

6 Too Much Pre-Heat

7 Too Little Pre-Heat

8 Tip Too Far From Plate

9 Tip Too Close to Plate

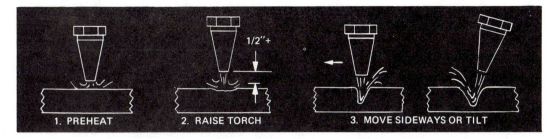

1. PREHEAT    2. RAISE TORCH    3. MOVE SIDEWAYS OR TILT

1/2"+

**Figure 4-19** Piercing to start cut away from edges. (*Courtesy of Airco*).

about 1/2" (12.7 mm) away and then depress the oxygen lever slowly to establish a hole (see No. 2, 3, and 4, Figure 4-19). After the hole is burned through, move the torch tip back to the normal distance away (see Figure 4-20). If the

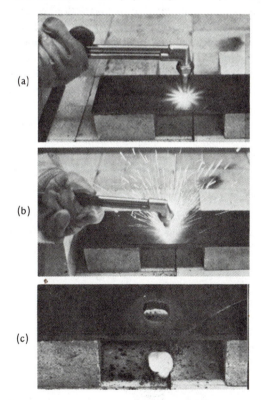

(a)

(b)

(c)

**Figure 4-20** Cutting a hole with the cutting torch: (a) preheating metal; (b) establishing hole; (c) cut to desired size. (*Courtesy of Victor Welding and Cutting.*)

tip is too close as the hole is being burned through, the slag will clog up the jets. The cutting torch is a real time-saver when used to make bolt holes—if the bolts do not require a close fit. With care and practice and using a bolt to check the size hole, it is much quicker than using a drill.

## BEVELING PLATES

With plates over 3/16" (4.76 mm) thick, the edges of the plates should be beveled in order to secure the proper penetration when welding the edges. When using a 4-preheat-jet tip, the line of travel should be between the two pair of jets (see Figure 4-7).

Using soapstone, mark a line near the edge of a plate. Light the torch and adjust the cutting flame. Starting from the edge of the cut, angle the torch about 45° with the plate. When the edge becomes a bright red, slowly depress the oxygen lever and watch how the cut is forming. Then, continue across the length of the plate. Observe the cutting marks and the smoothness of the cut. If it is difficult to hold the torch at a 45° angle, try using a piece of angle as a guide with the point of the angle facing upward, which will give a 45° angle on the sides (see Figure 4-21). In some instances where a lot of beveling must be done, and it is difficult to clamp the angle to the plate, a small piece of plate can be tack welded to the opposite angle; a clamp on the tacked pieces will hold the angle from moving (see Figure 4-22). Other angles can be made as in Figure 4-23.

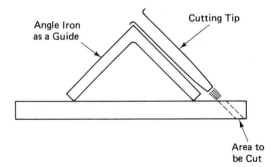

**Figure 4-21** Angle iron as a guide for beveling plates.

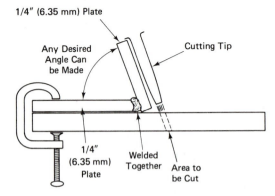

**Figure 4-22** Angle iron held in place with tab for beveling plates.

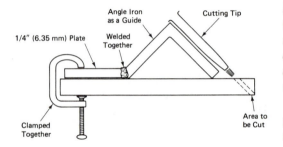

**Figure 4-23** Any angle can be made for beveling plates.

## ROUND BARS OR STOCK

Round stock or bars can be cut easily by holding the torch at 90° or vertical and starting at 90° from the top. As the cut is started, follow the contour of the bar, keeping the torch vertical to the surface. Close observation is needed to make sure the bar is being cut all the way through—especially when cutting through the thickest part (see Figure 4-24).

## CUTTING CAST IRON

Cast iron can be cut but it is a little harder than iron or steel. It takes longer to cut cast iron; it requires more heat and more oxygen. Also, the torch motion is different from regular cutting, and the cut will not be as smooth as before. There are many different types of cast iron; some will cut very easily while some of the poorer grades such as counterweights will be more difficult.

The oxygen pressure is about one-third greater for cutting cast iron than steel. In some cases, it is advisable to use one-half more oxygen. If the same thickness metal plate needs 30 psi (207 kPa), cast iron would require about 45 psi (310 kPa) of oxygen.

The torch setting is a little different with an excessive acetylene flame being used. 1/4"

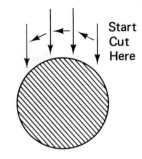

**Figure 4-24** Cutting round bar or stock. Keep torch vertical to the surface.

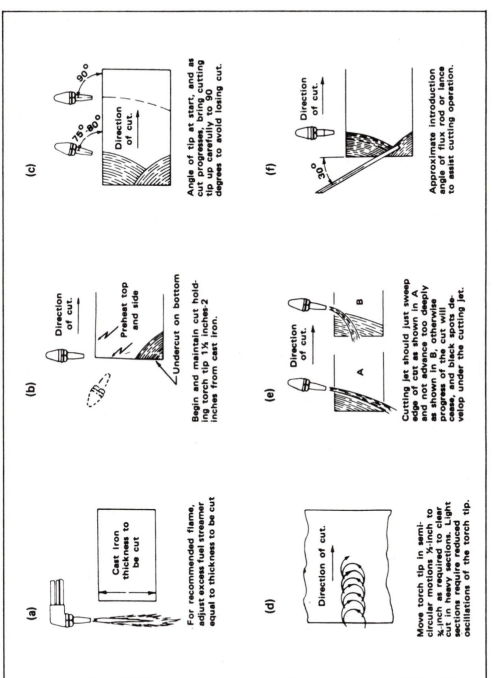

(a)

Cast iron thickness to be cut

For recommended flame, adjust excess fuel streamer equal to thickness to be cut.

(b)

Direction of cut.

Preheat top and side

Undercut on bottom

Begin and maintain cut holding torch tip 1¼ inches-2 inches from cast iron.

(c)

90°

75° 80°

Direction of cut.

Angle of tip at start, and as cut progresses, bring cutting tip up carefully to 90 degrees to avoid losing cut.

(d)

Direction of cut.

Move torch tip in semicircular motions ⅛-inch to ¾-inch as required to clear cut in heavy sections. Light sections require reduced oscillations of the torch tip.

(e)

Direction of cut.

A          B

Cutting jet should just sweep edge of cut as shown in A and not advance too deeply as shown in B, otherwise progress of the cut will cease, and black spots develop under the cutting jet.

(f)

Direction of cut.

30°

Approximate introduction angle of flux rod or lance to assist cutting operation.

Figure 4-25  Cast iron cutting. (*Courtesy of Airco.*)

55

to 1/2" acetylene feather is generally used for thin cast iron. For heavier pieces, a little longer feather is needed. The feather is determined when the cutting lever is depressed. The motion of the torch is a crescent motion rather than a straight movement of the torch. The kerf or cut will be a little wider than normal.

After adjusting the torch to an excessive acetylene flame, heat a small spot about 1/2" (12.7 mm) on the edge. Hold the torch at about a 45° angle to the edge. As the cast becomes hot, start to move the torch in a crescent motion and depress the cutting level. Once the cut has started, change the angle of the torch to about 75° (see Figure 4-25). Watch the speed of the torch. If it is too fast, the cutting operation will cease. On heavier pieces of cast iron, it may be very difficult to restart the cutting operation. If it is necessary to restart the cut, follow the same procedure as stated before.

## QUESTIONS

1. What are the two main objections to cutting with a torch?
2. What happens to the ferrous metal during the cutting process?
3. What are preheat jets or holes used for?
4. What is the center hole or jet in the cutting tip used for?
5. What is the average pressure used for cutting?
6. What is a kerf?
7. When cutting with a 4 preheat jet tip, how should the jet be aligned with the cut?
8. Before using a cutting torch, what must be done?
9. What are some of the personal safety precautions?
10. What about the welding hoses?
11. When using a cutting torch with 2 oxygen valves, how far should the torch valve be opened?
12. How hot should the metal be before starting to cut?
13. How much oxygen should be used or how far down should the oxygen be depressed for cutting?
14. What can be used to make a straight cut?
15. What can happen if the cutting tip is held too close to the metal when cutting a hole?
16. When bevelling plates with a 4 jet cutting tip, how should the jets be aligned?
17. When cutting round stock, how should the cutting torch be held?
18. Does cutting cast iron require more oxygen?

# BRAZING

## chapter 5

Brazing is another method of joining metal together. Brazing is very popular in the automotive trade and is also used widely for repair work in many other areas. Indeed, many repair jobs are ruined or fail becuase of the improper use of brazing materials or because of using brazing instead of welding. Perhaps some of the reasons for its popularity and widespread use are: the ease with which the process can be performed, especially for beginners or novices; the relative high strength of most of the joints; and the much lower temperature required as compared to steel welding. It requires less time and

skill to properly learn the brazing technique as compared to steel.

NOTE: The AWS designates brazing process as above 800°F (426°C) and soft solder as a process below 800°F (426°C). However, many repairmen and welders who are not familiar with the AWS desginations consider brazing as hard solder or a member of the solder family.

Brazing is defined as a group of welding processes where adhesion is produced by first heating the metal to a suitable temperature above 800°F (426°C) and then adding a non-ferrous filler that has a melting temperature below that of the base metal. (Non-ferrous means any metal that does not contain iron or steel.) The molten filler rod is distributed by capillary attraction between the closely fitted surfaces of the joint (see Figures 5-1 and 5-2).

## THE BRAZING PROCESS

Brazing is considered as hard solder or solder that is applied with a torch. Normally, when brazing is mentioned, it generally means the use of brass or bronze filler metal. But silver solder, some aluminum soldering processes, Phoscopper, and many more are in the same category of brazing.

Under the proper conditions, the filler metal will be distributed between the two closely fitted pieces of steel and other metals by a phenomenon by which adhesion between the molten filler metal and the base metal, together with surface tension of the molten filler, distribute the filler between the properly fitted surfaces of the joint to be brazed. This process is very similar to that of soldering copper pipes with wire solder which melts at approximately 400°F (204°C), and the copper at 1100°F (593°C).

When speaking of brazing, several common errors are made. One is that many welders use the term "bronze rod," which is copper and tin while talking about brazing, but they are really talking about brass rod, which is copper and zinc. Another refers to a repair job performed using the brazing method, for example, joining two pieces of heavy metal together, when it is

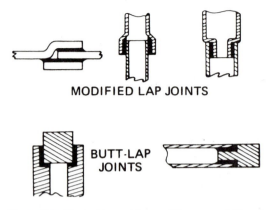

MODIFIED LAP JOINTS

BUTT-LAP
JOINTS

**Figure 5-2** Typical brazed joints. (*Courtesy of Airco.*)

really braze welding, which is described below (see Figure 5-3). One of the most common cast iron repair jobs involves beveling the broken or cracked edges and then filling the affected area with brass.

*BRAZE WELDING:* The filler metal is not distributed by capillary attraction. The process involves a groove, slot fillet, or plug weld, which is made by using a non-ferrous filler metal having a melting point below that of the base metal but over 800°F to 900°F (426°C to 481°C). This is shown in Figures 5-4 and 5-5.

In either case, brazing or braze welding, the base metal is not melted but merely brought up to a tinning temperature. If the base metal is permitted to melt, the base metal will mix with the filler metal causing a weak spot. Also, because the temperature far exceeds the normal brazing temperature, applying the brass rod to the surface will cause a flare up. This is caused by the decomposition of the zinc, noticeable by a trace of white powder in the area. This is the same white powder found where brazing or cutting galvanized steel.

Most brazed joints, when properly brazed, are very strong. In some cases they are as strong

Closely Fitted Pieces

**Figure 5-1** Brazing. (*Courtesy of Airco.*)

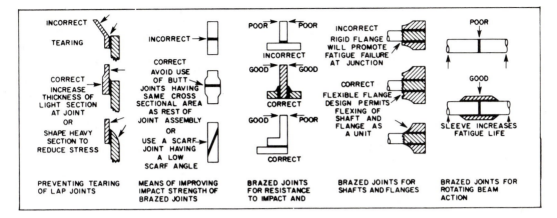

**Figure 5-3** Joint designs for brazing. (*Courtesy of Airco.*)

as steel welded joints. The weakest joint in the brazing process is the butt joint on light gauge metal. This joint should be avoided because of its frequent failure. In order to fix a broken brazed joint, the brass deposited on the base metal must be completely removed in order to reweld it with steel filler rod. Many inexperienced welders have made this mistake when trying to repair old fenders where cracks were caused by stress or vibrations.

When working with sheet metal or light gauge metal, heat can cause problems such as warping because of the large surface area of the panels. Because the metal is very thin, it reacts to any large amount of heat. This is why many repairmen prefer the brazing method over the steel welding method with its problem of heat. However, most of the brazing performed by the repairman would actually be braze welding.

Many exhaust systems can be easily repaired by the brazing process. The exception is butt

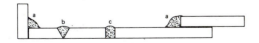

**Figure 5-4** Braze welding: (a) fillet; (b) groove; (c) plug.

welding two pipes together because of the constant expansion and contraction due to the exhaust heat. If this kind of repair is required, it is better to overlap the pipes and use the lap joint technique.

## BRAZING RODS

There are many different types of brazing rods available. The most common is the brass rod containing 70% copper and 30% zinc. This ratio will vary considerably according to the strength of the joint desired and the necessary temperature. Brazing rods can also contain small amounts of tin, manganese, iron, lead, or silver. A lower melting temperature can be obtained by adding tin and silver. Nickel bronze rods are made of copper and tin alloy with other compounds added for additional qualities or strength. Some manufacturers of brass or bronze rods will manufacture rods to your specifications. Brass and bronze are available either plain or flux covered; common sizes are from 1/16" to 1/8" (1.58 mm to 3.17 mm) in diameter and larger, and lengths of 18" or 36" (475.2 mm or 914.4 mm). Some repairmen prefer to use a bare brazing rod as a flux coated brazing rod can produce too much molten flux in the brazing area. This may interfere with the proper flow of the brass. Using powdered flux, the amount of flux used can be controlled by the welder.

## BRAZING FLUX

When steel welding, the base metal and the rod are melted and fused together. During the brazing process, the base metal is not melted and the brass will not adhere to the surface without the addition of a compound to clean the surface. A flux is needed to promote adhesion; it could be described as a material that removes oxides and film from the surface. A flux prevents oxides from forming while the metal is heated and permits the filler material to flow or wet the surface of the metal. The molten flux will also carry away impurities and prevent them from becoming imbedded in the brazed metal. Most of the popular fluxes contain borax or boric acid with small amounts of other chemicals such as bromides, fluorine, chlorine, and other compounds. The common household borax can be used for brazing but does not have as good a cleaning action as the regular flux. Specific brazing fluxes are made for different types of metals such as one for general repairs on steel or iron, one for cast iron, one for copper or brass, and many other types. Most are generally available in the powdered form (see Figure 5-6).

*CAUTION: Some fluxes will give off a very toxic fume when heated. Use them in a well-ventilated place. When removing the flux residue wear safety glasses because the residue will cause eye infection or damage. Remove flux from brazed joints because it may cause corrosion.*

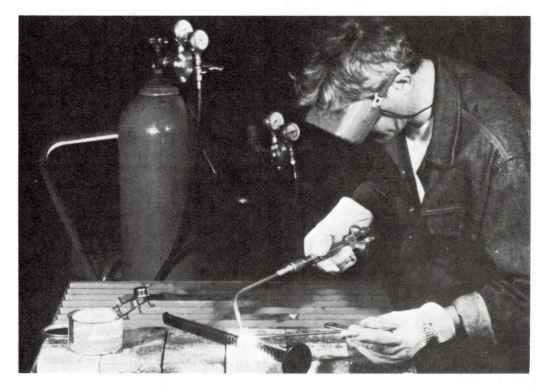

**Figure 5-5** Typical braze welding. (*Courtesy of Airco.*)

### BRAZO Flux (1-lb can)
- for braze-welding of steel, cast iron and malleable irons
- for fusion welding of copper, bronze and brass
- dissolves and fluxes away iron, copper, zinc and tin oxides

### OXWELD aluminum Flux (1/4-lb jar)
- for welding aluminum alloys
- lowers melting point of aluminum oxide to less than the molten base metal and floats the scale out of the weld zone

### OXWELD Cast Iron Brazing Flux (1-lb can)
- for cast iron braze-welding
- dissolves silicon and iron oxides
- tins base metal at brazing temperature

### FERRO Flux (1-lb can)
- for cast iron welding — prevents blowholes caused by inclusions
- dissolves silicon dioxide/iron oxide slag
- lowers melting point of slag for fluxing away at welding temperature

### CROMALOY Flux (1-lb can)
- for welding stainless irons and steels
- easily dissolves chromium oxide slags; removes other impurities and makes chromium alloys easy to weld

**Figure 5-6** Different types of flux. (*Courtesy of Linde Division of Union Carbide.*)

There are various opinions as to what type flame should be used for general purpose brazing, depending on one's background experience, or schooling. Generally, the most common practice is to use a neutral flame but the slightly reducing or carbuerizing flame, or a slightly oxidizing flame, are also used.

## BRAZING PROCEDURE

In discussions of the brazing procedures here, the procedure will be pointed more toward bodywork or repair work and working with 20- to 24-gauge (about 1 mm thick) sheet metal. Many of the brazing operations on the job will not be in ideal conditions. Most of the metal being brazed will be covered with paint, rust, undercoating, galvanizing, and other materials. This makes brazing more difficult; in addition, it is very difficult at times or too expensive to remove all the material covering the metal to be brazed. However, as much as possible should be removed by wire brushing, grinding, or scraping.

It is better for beginners to practice brazing under those conditions as soon as possible. At first, clean bare metal could be used to acquaint the beginner with the proper procedures and let him see the results. After this, the practice should be on old light gauge scrap that has traces of paint, rust, or other materials.

Use pieces of 22-gauge (about 1 mm) metal, about 2″ x 6″ (50.8 mm x 152.4 mm), a few clean and a few rusted or dirty pieces; 1/16″ or 3/32″ (1.58 mm or 4.76 mm) bare brass brazing rod; a #1 or #2 welding tip depending on torch make; brazing flux—powder type. Select a piece of metal and light the torch. Adjust the torch to a neutral flame. Bend over one end of the brazing rod for safety. Gently heat the end of the brazing rod and dip it into the can of flux. Notice that the flux will cling to the heated rod. Heat the metal to a dull red. Place the end of the rod on the heated area and melt off some flux. Immediately melt a small amount of rod. Notice if the brass flows freely on the surface; if it does, the metal is cleaned and the proper temperature has been reached. Add additional rod to build up a bead and advance the torch to make a bead. Notice how the molten flux will run ahead of the bead. Add additional flux to the rod when the molten flux no longer precedes the bead.

By proper torch manipulation, the bead will resemble that of steel welding as to contour and ripples. The width of the bead can be increased by adding more or using a heavier rod or by using an oval or weaving torch motion. If the bead is too high and the edges of the bead curve inward, the cause is either not

enough heat or not enough flux to properly clean the metal. Run several beads across the metal and check for uniformity in width, about 1/4″ (6.35 mm) wide, 1/16″ (1.58 mm) high and with uniform ripples. Next, try a dirty piece of metal without flux and notice the difference in flow as compared to using flux.

## BRAZING JOINTS

### DO NOT USE A BUTT JOINT FOR BRAZING—USE A LAP JOINT

If a situation arises in making some kind of repair where it seems a butt joint must be used, use a strip of metal under the joint for reinforcement—then it will be a lap joint. Experienced welders will confine their brazing to the lap and fillet joints in all positions. The lap joint brazing is frequently used when repairing body panels, light gauge metal, and others. Using the same method as used in steel welding to hold the panel in place (rivets, screws, clamps), the panel can be brazed easily. Due to the reduced amount of heat, the distortion of the panels will be less. With the reduction of heat, the seam or joint can be brazed completely thus making it watertight. Using asbestos along the joint and water to cool the bead after it has hardened and is no longer red will further reduce the chance of distortion. If the panel is handheld together while brazing, do not release the pressure until the brass has hardened or else it will crack.

Use a welding torch with #1 or #2 tip; two pieces of 22-gauge (about 1 mm) metal or scrap panels 2″ x 4″ (50.8 mm x 101.6 mm); 1/16″ or 3/32″ (1.58 mm or 4.76 mm) brazing rod; brazing flux; and vise grips.

CAUTION: Some fluxes will give off very toxic fumes when heated. Use them in a well-ventilated place. When removing flux residue, *wear safety glasses because the residue will cause eye infection or damage.*

Clamp the two pieces of metal together, one overlapping the other. Bend one end of the welding rod for safety. Set the torch to a neutral flame and heat the opposite end of the brazing rod, then dip it into the flux. Heat one end of the joint until it is dull red using more heat on the bottom pieces so they are equally heated. Then touch the flux-covered rod to the joint. When the flux has melted on the metal, add some rod along the edge of the joint. Notice the flow of the brass on both pieces; use enough brass for proper strength. The bead should be slightly concave with the top edge covered with brass and about a 3/16" (4.76 mm) wide bead on the bottom piece (see Figure 5-7). Proceed along the full length of the joint and notice the contour of the bead. Avoid adding too much brass and making the bead too big, as it wastes material. After it is cooled place the lower piece in a vise, bend the top piece apart, and look for signs of capillary attraction between the two pieces. Next, try the same procedure using dirty metal and check the results. If the brass spreads out too far, too much heat is probably the cause. If the brass bead is rounded, the cause is not enough heat.

Continuing on, next try strips of metal of different thicknesses remembering that the heavier piece or one with a greater surface area requires more heat. The bead or amount of brass deposited in the joint will vary with the thickness of the metal used.

## INSIDE CORNER OR FILLET JOINTS

Inside corner or fillet joints do not present a severe problem of distortion. The sufficient heating of both pieces is important—the larger area requires the most heat. The proper manipulation of the torch and adding the filler rod is important to give the necessary strength to the joint. The filler material should be distributed equally on both sides pieces; the resulting bead should be concave.

Take two pieces of 22-gauge (about 1 mm) metal or scrap light gauge metal, 2" x 4" (50.8 mm x 101.6 mm), welding torch with #1 or #2 tip, 1/16" (1.58 mm) or 3/32" (4.76 mm) brazing rod to practice.

*CAUTION: Some fluxes will give off a very toxic fume when heated. Use in a well-ventilated place. Wear safety glasses when chipping off the flux residue because the residue will cause eye infection or damage.*

Place one piece of metal on the other at a 90° angle (vertical) to the other. Support the vertical piece with a fire brick or other method. With the torch properly adjusted and the rod

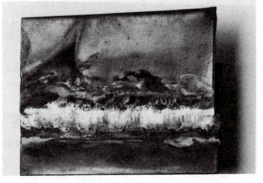

**Figure 5-7** A brazed lap joint. (*Photo S. Suydam.*)

**Figure 5-8** Brazed inside corner or fillet joint. (*Photo by S. Suydam.*)

fluxed, heat both pieces at one end to a dull red. Touch the flux on the metal; when melted, apply some rod. If the brass flows, add additional rod to the joint and proceed across the joint. Watch the bead formation and the flowing of brass on both sides. Check the bead for uniformity, width, and equal distribution of the filler material on both pieces. The bead should be slightly concave (see Figure 5-8).

Repeat this procedure using different thicknesses of metal.

## HORIZONTAL, VERTICAL, OR OVERHEAD BRAZED JOINTS

These joints will either be a lap or fillet joint when doing repair work. The biggest problem is gravity, and the improper use of heat will cause the brass filler to run downward. Avoid heating the metal beyond a dull red; using a large filler rod will help to eliminate the problem of the filler running down or away from the joint. Many of these joints are in a bad location where the welder is in danger of being burned from the filler that runs away from the joint. Inside the wheelhouse or on the bottom of the automobile generally present the biggest challenge to a repairman or bodyman. The use of flux-coated rods are recommended for these out-of-position joints because it may be difficult to properly locate the can of flux for easy access. The rod and torch positions are similar to those used for steel.

*CAUTION: Many shops use galvanized metal for repair work because it is easy to braze and in many cases, it can be obtained free as scrap. Use with caution and in a well-ventilated place. Avoid inhaling the fumes given off when heated, because they can cause serious illness.*

## ALUMINUM BRAZING

For many years, many welders shied away from repair involving aluminum. Today however, trends toward aluminum parts are increas-ing each year because of the weight factor and the conservation of energy. Now, several car manufacturers are using aluminum hoods on some of their automobiles with the forecast for additional use in the future. Perhaps the easiest way to weld aluminum is with the TIG welder or the MIG welder, but some bodymen and repairmen cannot afford to buy them.

NOTE: TIG and MIG welding are described further on in this text. Chapter 30 covers TIG Welders and the welding process. The MIG welder and the welding process are discussed in Chapter 31.

TIG (tungsten inert gas) welding is also called GTAW (gas tungsten arc welding), and MIG (metallic inert gas) welding is also called GMAW (gas metallic arc welding). Both welding processes involve the heat of the arc, with the arc being shielded by an inert gas. The high heat of the arc readily overcomes the high heat conductivity of the aluminum which makes these two processes excellent for welding aluminum, magnesium and other hard-to-weld metals. Both processes use rather complicated welding equipment.

Aluminum can be brazed or welded with the oxy-acetylene torch. Aluminum possesses several unusual properties, but once we understand them, the process becomes a little easier.

Aluminum can be brazed or welded in a way similar to iron. It shares with iron its great affinity for oxygen. Once aluminum is cleaned, it immediately reacts with oxygen to produce a thin, glass-like, hard film of aluminum oxide on the surface. This film of oxide gives aluminum its bad reputation. Immediately before the welding or brazing of aluminum, the oxides must be removed by scraping, wire brushing, sandpapering, or other means of mechanically removing the oxides. The oxides can be removed chemically by the use of a flux.

Another factor which makes aluminum different from steel is the high heat conductivity of aluminum. Aluminum conducts heat almost five times that of steel or iron depending on the type of alloy. This means that it takes more heat to bring aluminum to the proper welding or brazing temperature.

## PROCEDURE

Take a piece of steel sheet metal and a piece of aluminum, the same size and thickness, and heat one corner of both pieces for a few seconds. On the pieces of steel sheet metal, the heat will be concentrated at the corner while the heat will travel across the aluminum piece very rapidly. The biggest difference between steel and aluminum is the color change. When steel or iron is heated, it will gradually turn redder, then a bright red until it melts. Aluminum, on the other hand, will not change color until the melting point is reached and then it will simply fall away. One method used to determine the proper temperature to braze or weld aluminum is by using a special crayon that will melt or become liquid at a certain temperature.

**Flux** When brazing or soldering aluminum, a flux is used to obtain the proper adhesion. The flux is available in a powder or paste form. The flux should be applied to the surface to be welded or, if necessary, to the welding rod, especially if the rod has been in the shop for a period of time. The powdered flux can be mixed with water or as directed on the label.

**Aluminum Rods** There are many different types of aluminum rods available, the most common ones being the 1100 and the 4043. Some manufacturers produce a rod with a low melting point; it can be used for general aluminum repairs. When applied to the surface of the repair area, the flux will dry out as the temperature rises. As the correct temperature approaches, the flux will turn to a glossy liquid. Rub or scrape the aluminum rod in the heated area at 3 to 5 second intervals, removing the rod from the flame each time. When the right temperature is reached, the rod will flow. Keep the torch moving on and off the surface to avoid overheating. Avoid melting the rod with the torch before the surface is hot enough. The torch should be held at a flatter angle than when brazing or welding steel.

## WORKING WITH ALUMINUM

Aluminum has a tendency to become very soft around the weld area. This could lead to distortion and warping if the surrounding area is not supported by some means. This could be a major problem when repairing body panels that cannot be properly supported.

The field of aluminum welding, brazing, and soldering is very large because of the many different types of aluminum alloys and aluminum shapes such as tubing, forgings, sheets, and many more. Aluminum alloy may contain as many as six different elements to give it some desired property—being as hard as steel or almost as soft as lead. So the weldability of aluminum will change with each type of aluminum. When working with iron or steel, the type iron or steel can be approximately determined by the spark or chipping method. With aluminum, it is very hard to determine what type is involved. When working in the repair business, most of the parts or sheets have the code numbers removed or hidden so it is experience that helps determine what process is the best.

Cleaning the welded area after welding is very important because the flux is corrosive. If left on the aluminum, the flux will immediately cause oxidation or corrosion. Also, if the article is painted, any remaining flux residue will cause the paint to peel off. There are several different chemicals used by manufacturers to remove flux, but the use of boiling hot water and scrubbing with a brush will remove most of the flux. Then rinse the article with cold water.

*CAUTION: Aluminum flux may give off a toxic fume when heated. Use in a well-ventilated area. If necessary to mix powdered flux and apply to the surface, wash hands immediately and avoid getting flux in the eyes.*

The following procedures will be limited to brazing sheet and tubing, as most castings are very difficult to repair with a torch. Other methods such as TIG will give more satisfactory results and with less difficulty.

## BRAZING ALUMINUM

Use an oxy-acetylene welding torch outfit, several pieces of aluminum scrap 2″ X 4″ [(50.8 mm X 101.6 mm) gauge (1 mm), aluminum rod 4043 1/8″ X 36″ (3.17 X 9.14 mm)] or a rod of similar material, aluminum flux, small dish and acid brush, vise grips.

Remove oxides from brazing area by one of the above methods. Butt the two pieces of aluminum together, leaving about 1/32″ gap (.794 mm). Mix a small amount of the flux and brush a small amount of flux on the aluminum rod and on the joint area. Adjust the torch flame to a slightly excess acetylene flame or neutral flame. Heat one end of the butt with the blue cone about 1/8″ (3.17 mm) away, the torch about 20° to 30° with the joint. Watch as the flux turns to a powder then to a liquid. Touch the rod to the area, if the rod flows, the proper temperature has been reached. If not, apply more heat and move the torch to and from the area to avoid melting away the edges. As the rod starts to flow, keep the torch moving and dip the rod rather than keep feeding the rod steady. This will control the size bead. Continue across the joint.

This is a difficult method, but with practice it becomes easier. The biggest problem is melting away the edges. By moving the torch continually, there is less chance of melting the edges. Using a pair of pliers, scrub off the flux with a brush and hot water, then rinse in cold water. Observe the appearance and contour of the bead. Try this type again because aluminum

is more difficult than brass brazing, but it can be mastered.

## LAP JOINTS

Repeat the same steps as in the butt joints except for one difference. The bottom plate will require more heat than the top one and there is a danger of melting away of the edge of the top layer before the bottom has reached the proper brazing temperature. Tip the torch a little toward the bottom piece until the temperature is reached. Capillary action will carry the molten rod into the joint.

## INSIDE OR FILLET BRAZE

Using the same method as in the butt joint, keep the torch moving to avoid melting away the edges, especially the vertical one. With the proper temperature, capillary action will carry the molten rod into the joint.

## CAST IRON

Many objects made of cast iron can be easily repaired by braze welding. However, cast iron covers a large group of different kinds of cast. Likewise, many worn cast iron parts can be rebuilt by brazing. But brazing has its limitations, such as a part that has to stand a temperature over 500°F (255°C). Then brazing would not be recommended because the brazed metal would break down under that temperature. Also, a broken furnace grate would not be satisfactorily repaired by brazing.

Many objects repaired by brazing have failed because of the poor joint or surface preparation prior to brazing. Because actual fusion of the filler rod and the base metal does not take place (similar to brazing on sheet

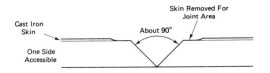

**Figure 5-9** One side accessible. Cast iron joint prepared for brazing using a single V-joint.

metal), other preparations of the base metal must be made.

Due to the nature of cast iron, merely removing the paint and other dirt or grease is not sufficient. The surface of cast iron along the edge of the break or worn spot must be removed by grinding, filing, or chipping with a chisel. The break must be ground or filed to a bevel forming a 90° angle if only the area side is accessible. If both sides of the break are accessible, the break area should be beveled on both sides for additional strength.

*CAUTION: This flux, when heated, gives off fumes which may irritate eyes, nose, and throat. Use in a well-ventilated area. Contains fluorides.*

Use cast iron scrap, an oxy-acetylene welding outfit with suitable tip, 1/8" (3.17 mm) brazing rod for cast iron, suitable flux for cast iron, and fire brick. Select a piece of cast iron, preferably pieces with matching break. Remove all rust, oil, or other foreign matter. If only one side is accessible, bevel the edges of the break to form a 90° bevel all the way through the thickness of the casting (see Figures 5-9 and

5-10). Remove the surface coating about 1/2" (12.7 mm) along the full length of the break.

NOTE: A 90° angle would be the combined angle of both pieces were each beveled to a 45° angle. Also, the more surface area in the break area, the greater the strength of the repair.

Align the two pieces together on the fire brick; preheat the pieces until the ground areas start to turn dark. Heat the brazing rod in the flux, and using a neutral to slightly oxidizing flame, tack braze the two pieces until the ground areas stay in place. Start from one end and braze the length of the joint. If the metal is heavy, do not attempt to fill in the groove in one pass. It is better to make two passes or three if necessary. Be sure that the braze rod is wet or tins the complete surface of the groove. Inspect the completed joint, looking for spots where the braze did not flow out on the surface (see Figure 5-11). The bead should be slightly convex. Repeat this procedure using different types of joint designs such as those used on a cast iron frying pan handle or an object where the break can be brazed on both sides.

*CAUTION: Never, at any time, quench or cool with water a piece of cast iron that was just brazed. The sudden cooling could crack the cast. Let it cool as slowly as possible. If possible, cover it with dry sand or powdered asbestos.*

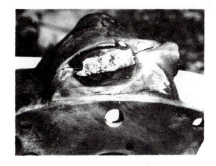

**Figure 5-11** Brazed cast iron joint.

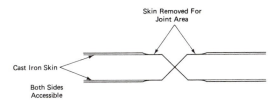

**Figure 5-10** Both sides accessible. Cast iron joint prepared for brazing using a double V-joint.

## QUESTIONS

1. What makes brazing very popular?

2. What type filler rod is used most generally for brazing steel?

3. How is the filler metal distributed between closely fitted surfaces of a joint?

4. Brazing is used with what type solder?

5. What is brazing welding?

6. How does brazing differ from steel welding?

7. What causes a flare-up when brazing?

8. What type joint is not recommended when brazing thin-gauge metal?

9. What are brass brazing rods made of?

10. What is the reason for using brazing flux when brazing?

11. What is the advantage of using powdered brazing flux?

12. What is the purpose of using dirty or rusted pieces of metal when practicing brazing?

13. How hot should the metal be for brazing?

14. What causes the edges of the bead to curve inward?

15. What can be used to control distortion when brazing light-gauge metal?

16. What type flame is generally used for brazing?

17. When brazing a lap joint, where should most of the heat be applied?

18. What problem exists when brazing in a vertical position?

19. What are some characteristics of aluminum that makes it more difficult to weld or braze?

20. What are some methods to remove the oxide from the surface?

21. Where should flux be applied?

22. Why is post-cleaning necessary after repairing aluminum?

23. Why are lap joints different in welding or brazing aluminum?

24. How should cast iron be prepared for brazing?

# ARC WELDING

## chapter 6

This text is designed to help the inexperienced welder gain more knowledge in the use of the Arc Welder. For some, it will be their first exposure to the field of electric arc welding or (SMAW) Shielded Metal Arc Welding. Shielded Metal Arc Welding is a small part of the vast field of fusing metal by use of the electric current. Yet it is the most widely used process in industrial production, repairing heavy equipment, farm machinery, and transportation vehicles and also most popular with the "at home" hobbyist. With the increased popularity of the low cost AC welder, more and more AC welders are finding their way into many backyard garages and small shops. Many objects can

be repaired or rebuilt thus avoiding costly replacements or time lost waiting for parts.

Many of the procedures will vary from other previously published books on Arc Welding, but these concepts were developed by the author after spending many years in the repair field. It is up to the beginner to develop his own technique—what works best for him may not work for someone else.

Welding is a hand trade in which experience is the only and best teacher. The technical knowledge is extremely useful in understanding the various parts of the trade, but the ability to transfer this knowledge into useful work is the main goal. Throughout this unit, there will be many references to actual repair jobs or problems that might arise when involved in the repair trade.

## THE DEVELOPMENT OF ARC WELDING

The art of welding two pieces of metal together is an ancient craft, but the science of arc welding is fairly new, less than 200 years old. Many nations contributed to the development of the arc welder. Back in the early 1800s, an Englishman, Sir Davy, discovered that an electric current would jump an air gap. It was not until much later in the 1800s that arc welding was first practiced. A Frenchman, DeMeritians, joined storage battery plates together with a carbon arc. Later on, a Russian discovered that the heat of the arc would melt a bare metal rod and the rod would act as a filler metal in a weld. This rod was given the name, "electrode."

The electrode wire or rod, which can be bare or flux coated and which has the same approximate chemical composition as the metal to be welded, carries or brings the current up to the point where the arc is to be formed. The arc is formed between the ends of the electrode and the base metal, which melts off the end of the electrode and the base metal. The molten metal of the electrode flows into a crater formed in the molten base metal and fuses together either to form a permanent bond between the pieces of metal or to form a bead.

These bare electrodes were very difficult to use, and the welds were weak and porous (see Figure 6-1). American patents on the metal arc process were issued to Charles Coffin in 1889. In 1908, a Swede, Oscar Kjellborg, experimented with applying a coating to the bare electrode. This made the welds stronger because the chemicals contained in the flux cleaned the metal

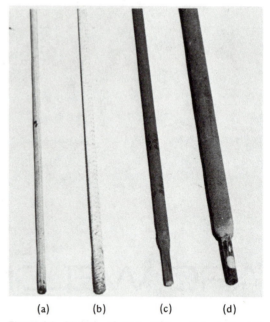

(a)      (b)      (c)      (d)

**Figure 6-1** Picture of bare and coated electrodes: (a) & (b) Bare electrodes; (c) Thin coated electrode (E6011); (d) Heavily coated electrode (E7024).

and aided in stabilizing the arc. Kjellborg's development is responsible for an estimated 90% of the present day manual arc welding. It was not until about 1927 that the development of the mass-produced extended type coated electrode occurred, which brought down the cost of electrodes. With the outbreak of World War I, the need for increased production of war machines and their repairs gave the arc welding process a chance to prove its capabilities. In the years following, it was used to manufacture machinery and storage tanks, build bridges and pipelines (see Figures 6-2 and 6-3), and accomplish repair work (see Figure 6-4). During this time, the manufacture of the welding machines vastly improved. At the start of World War II, the arc welder proved itself by the inexpensive and rapid fabrication in the mass production of merchant ships, tanks, aircraft, and the many

Figure 6-2 Fabrication of large structures. (*Courtesy of Lincoln Electric*.)

Figure 6-4 Repairing machinery in the field. (*Courtesy of Lincoln Electric*.)

indispensable tools in the industrial world. The welded joint replaced the costly riveted joint (see Figure 6-5). Many castings in the machinery were replaced by welded steel shapes with great savings in weight and increase in strength. This led to simpler designs, faster fabrication, and saving time and material. Technology is still making improvements in the machines and the electrodes. The simple transformer type welder, most commonly used today, weighs a fraction of the ones made

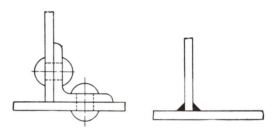

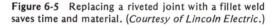

Figure 6-5 Replacing a riveted joint with a fillet weld saves time and material. (*Courtesy of Lincoln Electric*.)

Figure 6-3 Manufacture of construction equipment. (*Courtesy of Lincoln Electric*.)

Figure 6-6 Transformer type welder. It is many times lighter than the older models. (*Courtesy of Miller Electric*.)

thirty years ago (see Figure 6-6). Electrodes are made to weld almost every alloy in existence today and with increasing strength of the welded joint.

## FUNDAMENTALS OF THE ELECTRIC CURRENT

Electric Arc Welding is defined by the American Welding Society as follows: "A group of welding processes wherein coalescence (to unite or combine) is produced by electric heating with an electric arc or arcs with or without the application of pressure and with or without filler metal."

Because arc welding deals with electricity, let us define it. Electricity is described in the dictionary as follows:

A fundamental physical agency caused by the presence and motion of electrons, protons, and other charged particles, manifesting itself in attraction, repulsion, magnetic, luminous and heating effects and the like.

The following list of words and definitions will be used many times throughout this text:

Unit of electrical pressure—called volt

Unit of electrical quantity—called ampere

Unit of electrical resistance—called ohm

Current means the flowing of electricity

Static means stationary electricity

Conductor—a material that will allow the flow of current

Non-conductor—permits little or no flow of current, also called insulator or di-electrics

Di-electrics is also an insulation between conductors of opposites. This term is generally used only when induction takes place through it.

Direct Current (DC)—electric current flows only in one direction

Alternating Current (AC)—electrical current that reverses in direction periodically or 120 times a minute and completes 60 cycles per minute

Polarity is the direction of the flow of current mainly used in DC welders

## QUESTIONS

1. About how old is arc welding?

2. What were the first types of electrode used?

3. When were coated electrodes first mass-produced? What effect did it have on their price?

4. How does the transformer type welder of today compare with the one used 30 years ago?

5. What is a unit of electrical pressure called?

6. What is a unit of electrical quantity called?

7. What does current mean?

8. What is direct current?

9. What is polarity?

10. What is alternating current?

# SAFETY IN ARC WELDING

## chapter 7

Before a person uses the arc welder for the first time, he should be aware of the potential hazards of the trade. By using common sense and observing the safety rules, arc welding can be a safe occupation. There are several things in arc welding that can cause serious injury if the welder is careless: the high electricity involved; the ultraviolet and infrared light rays given off by the arc; the molten metal, sparks, and molten metal globules that fly from the arc; the handling of or presence of hot metal; and the fumes given off are just some of the things that can cause injuries.

## PERSONAL SAFETY MEASURES

### THE WELDING HELMET

The welding helmet is a must when arc welding to protect the welder's eyes from being burned by the ultraviolet and infrared rays given off by the electric arc. The dark lens in the welding helmet will filter out about 99% of the harmful rays. The arc rays are considered harmful even over 50 ft. (15.2 m) from the source. Also, the helmet protects the welder's face and neck from the rays and hot sparks coming from the arc (see Figure 7-1). Looking at the arc at a close distance for a few seconds can cause serious injury to the eyes; prolonged exposure to the rays can cause blindness. The rays will also cause serious skin burns to the face and neck if they are not covered by the helmet. Dark sunglasses or gas welding goggles must never be used because they will not give sufficient protection. The helmets are equipped first with an inexpensive clear lens in front to protect the filter lens from flying sparks. The clear lens is changed frequently when it becomes covered with spatter and visibility is impaired (see Figure 7-2). Immediately behind but separated by a fiber gasket is the filter lens,

which comes in different shades. Shade #10 is used for small electrodes, #12 for larger electrodes, up to #14 for use with the carbon arc. The higher the number, the darker the lens. Behind the filter lens is another clear lens and a fiber gasket to protect the filter lens. Never, at any time, use a helmet with a cracked filter lens or one without a clear lens in front of the filter. Small chips out of the filter lens can reduce the amount of rays filtered out. Before using any welding helmet, check the condition of the helmet and lens.

NOTE: Magnifier lenses are available for welders with sight problems or who have trouble welding with bi-focal eye glasses (see Figure 7-3).

## CLOTHING

Shirts made of heavy material, preferably wool, but free of oil, grease or other flammable liquids, with long sleeves, offer a safe amount of protection if leather clothing is not available (see Figure 7-4). To avoid the reflection of rays underneath the welding helmet, wear dark, colored shirts. Avoid fuzzy sweaters, new flannel shirts, and open pockets—especially with matches or combs in the pockets.

NOTE: Butane cigarette lighters should be removed from the pockets and taken away from the welding area. These lighters have been known to explode with considerable force when under certain heated conditions.

Avoid short-sleeved summer shirts or shirts of man-made materials. Short-sleeved summer

**Figure 7-1** Helmet with rectangle and square lens or plates. Some prefer one to another. (*Courtesy of Airco.*)

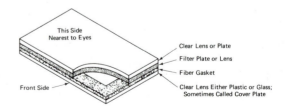

This Side Nearest to Eyes

Front Side

Clear Lens or Plate
Filter Plate or Lens
Fiber Gasket
Clear Lens Either Plastic or Glass; Sometimes Called Cover Plate

**Figure 7-2** Welding lens assembly.

---

**HOW TO SELECT WELDING MAGNIFIER**

| Age | Magnifier Focal No.* | Age | Magnifier Focal No.* |
|---|---|---|---|
| Up to 44 | 1.00 | 51-56 | 2.00 |
| 45-50 | 1.50 | 57 and Over | 2.50 |

*Based upon 18″ normal working distance.

---

## HOW MAGNIFIER HELPS WELDER

**WITHOUT**

Bifocal vision through helmet window is limited by improper alignment of window and bifocal segment. Wearer cannot use large upper portion of his glasses.

**WITH**

MAGNIFIER lens now in holder behind welding plate, bifocal power is provided through entire window area, with near point vision restored . . . morale and production improved.

Magnifiers are individually boxed, with focal numbers clearly identified.

Drawing (above) shows how Magnifier lens for each eye is optically centered, resulting in improved vision.

**Figure 7-3** Welding magnified. (*Courtesy of Linde Division of Union Carbide.*)

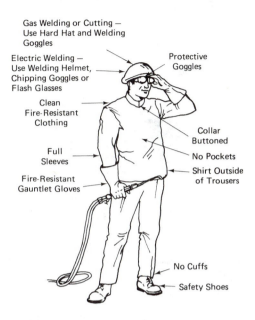

Gas Welding or Cutting — Use Hard Hat and Welding Goggles

Electric Welding — Use Welding Helmet, Chipping Goggles or Flash Glasses

Protective Goggles

Clean Fire-Resistant Clothing

Collar Buttoned

Full Sleeves

No Pockets

Shirt Outside of Trousers

Fire-Resistant Gauntlet Gloves

No Cuffs

Safety Shoes

**Figure 7-4** When gas welding or electric welding, select clothing to provide maximum protection from sparks, hot metal and harmful rays. (*Courtesy of Linde Division of Union Carbide.*)

shirts leave the arms, chest, and other parts exposed to the harmful rays causing sunburn, similar to being exposed to the sun but a more rapid burn because of the closeness to the source of light. After persistent exposure to the rays, over a period of time, a skin condition similar to sunpoisoning could occur.

Many of the man-made materials or synthetics will melt before they burn and will adhere to the skin, causing painful burns.

Pants should be of the work type and not of man-made material, without cuffs because cuffs serve as spark catchers; pants should be long enough to cover the tops of the shoes. Pants with frayed bottoms are a hazard.

Shoes of the work-type with high tops, preferably with steel toes, are recommended. Sneakers, low oxfords, or man-made fiber shoes are dangerous because of the hot metal and the heavy metal being handled.

NOTE: Man-made fiber shoes will melt when in contact with hot metal.

Work gloves with cuffs or gauntlet type gloves should be worn, but some welders find it hard to work with gloves—they have to be extra careful to avoid burning their hands.

Some industrial shops require protective clothing such as leather aprons, sleeves, anklets, gloves, steel-toed shoes, and skull caps be worn, but the average small shop has little or nothing in the line of protective clothing (see Figure 7-5).

## SAFETY GLASSES OR CHIPPING GOGGLES

Safety glasses must be worn when chipping the slag from the weld or when handling freshly welded metal because of the hot slag or globules of metal in the welded areas. They can cause serious eye damage when hot; even when cold, they contain oxides that cause eye infection (see Figure 7-6).

**Figure 7-5** Welding jacket and gloves. (*Courtesy of Hobart Brothers.*)

## MAKING THE ENVIRONMENT SAFE

### PROPER VENTILATION

When welding, large amounts of smoke, gases or fumes, and dust are given off in the weld area either from the electrode coating or the metal being welded. The metal may be covered with paint or other materials. The smoke and fumes may be toxic and could cause injury to the respiratory system. Some electrodes are worse than others, so weld in a well-ventilated area.

If the shop lacks an exhaust system, and the work cannot be moved, a regular household fan will disperse the fumes or blow the fumes away from the welder. Many times, the welder must use his own ingenuity for his personal safety, especially in a small shop (see Figure 7-7).

*CAUTION: When welding on metals that contain or are coated with zinc (galvanized), lead, cadmium, beryllium, or mercury, positive ventilation is a must for the welder's safety.*

It is recommended that a hose mask, which supplies clean air, or self-contained breathing equipment or other devices recommended by the U.S. Bureau of Mines, be used if the welder is being exposed to large amounts of metallic fumes.

Whenever a welder works in a confined area, such as inside tanks or metal containers, an adequate supply of air is necessary for breathing and a means of exhausting any accumulations of toxic gases or fumes is also necessary (see Figure 7-8). A hose mask is recommended and also a person or an assistant on the outside to watch out for the safety of the person inside. Welding on metal covered with a degreasing compound or with chemicals that may decompose with heat and the rays from the weld area could cause irritation to the eyes and respiratory system. Before welding, check the area for ventilation or set up some means of

**Figure 7-6** Chipping and grinding goggles. (*Courtesy of Linde Division of Union Carbide.*)

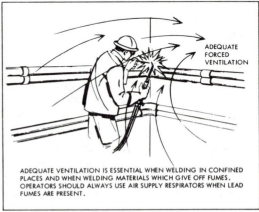

ADEQUATE FORCED VENTILATION

ADEQUATE VENTILATION IS ESSENTIAL WHEN WELDING IN CONFINED PLACES AND WHEN WELDING MATERIALS WHICH GIVE OFF FUMES. OPERATORS SHOULD ALWAYS USE AIR SUPPLY RESPIRATORS WHEN LEAD FUMES ARE PRESENT.

**Figure 7-7** Adequate ventilation. (*Courtesy of Linde Division of Union Carbide.*)

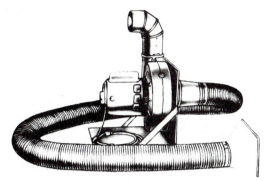

**Figure 7-8** Use an exhaust system to exhaust toxic fumes when welding in a confined area. (*Courtesy of Hobart Brothers*.)

DON'T weld on hollow (cored) castings that have not been properly vented. The casting may explode.

**Figure 7-9** Don't weld on hollow (cored) castings. (*Courtesy of Airco*.)

ventilating the area where the welding is to take place.

NOTE: Use caution when welding on used containers such as drums, tanks, and other types unless they are thoroughly cleaned. Depending on what type of material was stored in the containers originally, dangerous toxic fumes or explosive gases could be given off as the metal is heated or welded.

*CAUTION: Never weld or heat a sealed or closed container (see Figure 7-9). The container should be adequately vented. Fill with water to just below the weld area to avoid accumulation of gas and fumes, if necessary (see Figure 7-10).*

## SURROUNDING AREA

In many shall shops there is no special place for welding. Therefore the welder, before welding, must check the area for combustible materials such as oily rags, paint, or anything that will burn. The combustibles must be moved away from the area, and a fire extinguisher kept handy. Some shops have a pail of water and a pail of dry sand in the welding area. When welding on trucks, check the location of the fuel tank and cover it with an asbestos blanket (if available) or cover it with a water-soaked blanket. Use spark shields on the floor area to keep the sparks from traveling.

*CAUTION: It is very easy to have a fire start alongside the welder without his being aware of it because of his limited vision resulting from his helmet. If a ventilating fan is operating, it will draw the smoke away and the welder will not see or smell it. A few minutes spent in clearing the area of combustible materials will save hours later on.*

## WELDING OR GENERAL SHOP AREA

If it is possible, provide a welding screen or curtain between the arc welder and the other employees or students working in the shop.

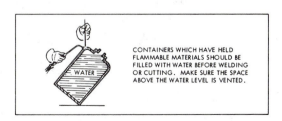

CONTAINERS WHICH HAVE HELD FLAMMABLE MATERIALS SHOULD BE FILLED WITH WATER BEFORE WELDING OR CUTTING. MAKE SURE THE SPACE ABOVE THE WATER LEVEL IS VENTED.

WATER

**Figure 7-10** Welding on containers. (*Courtesy of Linde Division of Union Carbide*.)

Many times this is impossible, especially if there are several welding machines being operated at one time, or if the welding is being done on a high piece of machinery or a truck or trailer.

All persons working in the shop or visitors should be warned not to look at the light of the arc without proper eye protection. Some shops provide flash glasses to the people working near the welders to reduce the danger of eye damage.

## WET FLOOR

It is recommended to weld only where the floors are dry. Many times, in the repair field, welding must be done on equipment that is wet or where the floor is damp. If the cables are in good condition, electrode holder well insulated, the ground clamp, and the machine properly grounded, there is little danger of severe shock. Good rubber boots and gloves will help in this situation. Emergency repairs must be done, rain or shine.

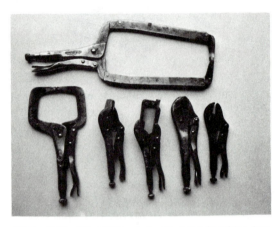

**Figure 7-11** Vise-grips can be used for handling hot metal. (*Courtesy of Airco.*)

**Figure 7-13** Don't use worn or poorly connected cables. (*Courtesy of Airco.*)

**Figure 7-12** Never pick up hot objects. (*Courtesy of Airco.*)

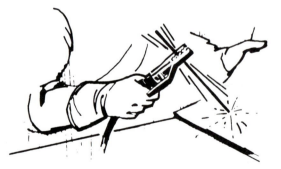

**Figure 7-14** Don't use electrode holder with defective jaws. (*Courtesy of Airco.*)

**Figure 7-15** Don't overload cables. (*Courtesy of Airco.*)

## SAFETY CHECKS FOR WORKING

### *HOT METAL*

Metal will be very hot after welding. Wearing gloves and using tongs, vise grips, or pliers are recommended (see Figure 7-11). Avoid sudden cooling of the metal because it can weaken or fracture the metal or weld. This could cause it to fracture or crack. In some cases, internal stress in the metal could cause failure later on. If it is necessary to leave the area after welding, mark the metal HOT with soapstone so that someone else will not touch it (see Figure 7-12). If possible, place the hot metal under the bench or someplace out of the way. This should also be done to metal recently cut with a cutting torch.

### *CHECK MACHINES AND CABLES*

Before turning on the machine to start welding, check over the welding cables for bad spots in the insulation (see Figure 7-13). Check the electrode holder for loose jaws, defective insulated handles, or set screws for the cable protruding through the insulation (see Figure 7-14). Check the ground clamp for positive grip jaws and a tight cable connection on the clamp. Poor grounds, or a bad connection in the electrode holder or ground clamps can cause overheating and loss of current (see Figure 7-15).

## QUESTIONS

1. What are some of the dangers in arc welding?
2. Why is a welding helmet necessary, besides protecting the eyes?
3. What percentage of rays are filtered out by the welding lens?
4. Why should a clear lens be used in front and back of the dark lens?
5. What type shirt should be worn when arc welding?
6. What is the danger in wearing light summer shirts?
7. Why should pants with cuffs be avoided?
8. What type of shoes is recommended?
9. Should chipping or safety glasses be worn when chipping slag?
10. Why is it important to weld in a well-ventilated area?
11. What are some types of metal that should be welded with extreme caution?

12. What should be done before welding on used containers?

13. What should be done to hot metal left lying on a bench?

14. Is it safe to weld on sealed containers?

15. What should be done before starting to arc weld?

16. If the welding area cannot be screened off, what should be done?

17. Name a few things that should be checked before using the welding equipment.

# AC AND DC WELDING MACHINES

## chapter 8

Arc welding uses the heat produced by the electric current jumping the air gap from the electrode to the work. A very concentrated and intense heat is produced ranging up to about 7000°F (5500°C). The power supply or welding machines needed to produce the necessary current to jump the air gap are classified either AC or DC (alternating current or direct current). Some machines are made to produce both AC and DC current.

## AC WELDING MACHINES

The AC machines are increasing in popularity. The AC machines now being produced are very efficient and easy to handle (see Figure 8-1). A large selection of electrodes has been developed for use with AC machines; these electrodes have an ionizing compound in the electrode coating. The ionized arc stream makes it easier to maintain a stable arc and to strike an

**Figure 8-1** A very popular AC welder for small shops. A limited input transformer with a dial type control. (*Courtesy of Lincoln Electric.*)

arc. The original cost of the transformer AC type is less than the rectifier type or motor generator of the same equal welding capacity. With AC current, it is easy to step up or step down (increase or decrease) the voltage by the use of transformers. Most AC welders that step down the voltage increase the current for welding. The input is directly related to the output.

The output of AC welders ranges from 150 amps (see Figure 8-2), used for light duty welding, to 1500 amps (see Figure 8-3), for heavy-duty production welding. All welders (machines) should comply with the National Electrical Manufacturers Association standards. The Association is concerned with the allowable temperature rise in parts of the machine as it is

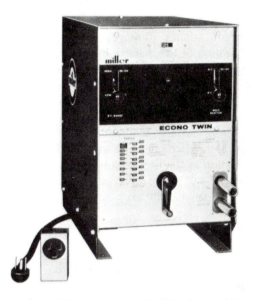

**Figure 8-2** 150 ampere output for light duty welding. (*Courtesy of Miller Electric.*)

**Figure 8-3** 1875 ampere output for heavy duty continuous welding. (*Courtesy of Miller Electric.*)

operated at its rated output. Temperature rise maximums are based on the type of insulation material used. All machines, either AC or DC, should be manufactured with safety in mind.

However, some machines, especially the less expensive or lower amperage (100 amp, 110 volts) should be avoided because they may not be up to standard. It is recommended that the open circuit potential not exceed 75 to 80 volts for safety factors.

Open circuit potential or voltage is the voltage produced when the machine is running and no welding is being done—sometimes called idling. The actual voltage used to provide the proper welding current varies from 15 to 40 volts. (The approximate input for a given welder could be 225 volts at 25 amps with an output of 30 volts and 200 amps.) When the arc is struck, the voltage drops down between 15 to 40 volts. This higher voltage makes it easier to strike the arc with most electrodes.

## ADVANTAGES OF THE AC WELDING MACHINES

The AC transformer welder provides lower cost operation and is practically maintenance free with no moving parts except for the cooling fan. There is an absence of arc blow or it is kept to a minimum because the current constantly changes direction. (Arc blow will be discussed in the DC machine section.)

Because of the absence of arc blow, heavier gauge steel can be welded using higher current and heavier electrodes, which makes AC welding faster. With the use of ionizing compounds in the electrode coatings, the AC welder has a good, forceful arc. The AC welders, combined with a high frequency unit, are excellent for welding aluminum.

## DISADVANTAGES OF THE AC WELDING MACHINES

The disadvantage of AC welders is that they are limited in some applications to certain types of welds. Some electrodes are made to be used on DC machines only. One of the problems with the transformer type welder is that the output voltage is affected by the input voltage or power source. In some shops or buildings, the power supply is inadequate to run all of the equipment, or the line voltage is low or fluctuates. This has a definite effect on the welder output.

## OPERATION OF THE AC MACHINE

In the AC current, the alternating current reverses its direction of flow 120 times per second. Therefore, it requires 1/60 of a second to complete the cycle and completes 60 cycles per second. The Sine Chart, Figure 8-4, shows the cycle: the voltage at A and B is zero, then the voltage builds up to C, the maximum in one direction. Then back to zero at point A, then to the maximum at D in the other direction, re-

turning B to start the cycle over again. As the current cannot flow at zero voltage, the ionization of the material in the electrode coating helps to maintain the arc flow across the air gap. This was the important factor in the growth of the AC welding process. With the current reversing itself the same number of electrons would theoretically travel in one direction in the arc and then in the opposite direction as the flow of current reverses. Therefore, 50% of the heat is released at the electrode and the same at the work side.

### OPERATION

Current enters the primary windings, which are wrapped around a laminated iron core (see Figure 8-5). The magnetic strength of the coil is directly proportional to the number of wind-

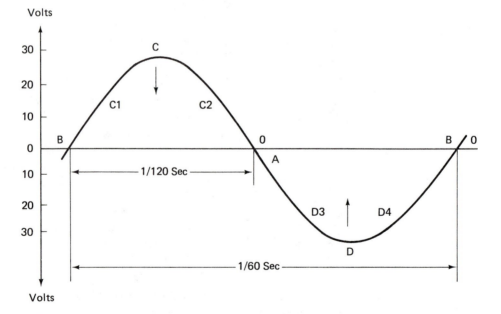

**Figure 8-4** Sine wave of alternating current at line A-B, the voltage is 0, maximum voltage at C and D. Theoretically, the current will jump the gap at C1, reach the maximum at C and start to decrease as it passes C2 and is extinguished at 0 (A). The arc will start again at D3, and reach the maximum at D, then decreases at D4, until extinguished at 0 (B). This sequence happens 60 times a second.

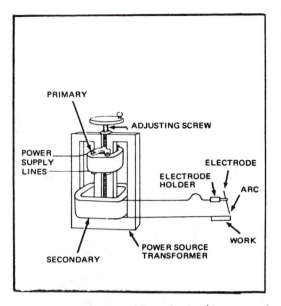

**Figure 8-5** A simple welding circuit. (*Courtesy of Airco Welding.*)

## TYPES OF AC MACHINES

Different types of devices are used to select the proper amperage or current. Several commonly used current controls will be discussed in this section.

One of the simplest methods is the use of a selector dial or a rotating switch. The center of the switch changes taps, or leads, from the secondary windings, which in turn change the current. On observing the back of the selector dial or rotating switch, one can see that the leads or taps of the secondary windings are much smaller or lighter for the lower current setting. For example, the taps will be 40 and 60 amperes as compared to the much heavier 180 or 225 amperes leads. However, the selector dial method is limited in its range of ampere or current selection. The range on many of these machines is, for example, in steps from 30 amp, 45 amp, 60 amp, 75 amp and so on, up to the maximum range of the machine which is usually 180 or 225 amperes. This limitation sometimes creates a problem if 75 amperes is not hot enough but 90 amperes is too hot. The amperes control the amount of heat available in the arc stream. The higher the ampere setting, the greater the heat.

Another type of current control is the tap-in or tapped reactor, where the electrode lead is tapped (see Figure 8-6), or plugged into the different taps. Some machines use a high, medium, and low ground tap which provides a wider range of amperage or current selection than the single ground tap-in or the previously discussed selector dial type machine.

The tap-in welders use a series of tap sockets which are attached to different places in the secondary windup. Likewise, in machines with different ground taps, the ground taps are attached at different places on the ground side of the secondary windings. Using Figure 8-7

ings or turns of the coil. These windings are insulated from each other and the core. When the magnetic field from the primary windings moves back and forth across the secondary windings, current is produced. The secondary windings are heavier and fewer in number. The welding leads are connected to the secondary windings. Some machines have capacitors to improve the power factor by discharging and increasing the wattage as the sine wave of the input power decreases from the maximum and approaches 0.

On some machines, additional primary windings and capacitors are added to improve the power factor. When approaching 0, the additional third coil releases electrons, thus producing a stronger magnetic field for a longer time. Reactor coils are also used in some machines. They delay the rate of changes of current and store electromagnetic energy. They also add to the arc stability, counteracting any influence which tends to extinguish the arc.

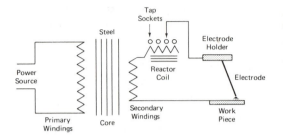

**Figure 8-6** Basic AC arc welding transformer. Tap in sockets with reactor coil with no moving parts. Note: The reactor coil is used to provide a more steady current.

as an example, the electrode lead plugs into a 90, 100, and 110 ampere tap. The 90 ampere setting can be obtained by plugging the ground cable into the low ground tap or the 110 setting can be obtained by plugging the ground into the high ground tap.

Some AC machines have a movable core (see Figure 8-8), which is operated by a crank handle that moves the steel in or out of the windings. To increase the current, the core is moved inward; and the current is decreased by moving the core out. This machine gives the

**Figure 8-7** AC welder with tap-in for selecting the proper current electrode cable tap in on upper scale ground cable in either hi, med or low. (*Courtesy of Marquette Division of Applied Power.*)

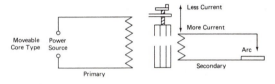

**Figure 8-8** Different types of AC welding machines.

operator a larger range of welding amperes.

Another type AC machine has a movable coil (see Figure 8-9), where the coil is moved to or away from the iron core. By moving it in, the current is increased; moving the core out decreases the current flow. The rheostat type has a resistance coil used to control the flow of current. This machine also permits a wide selection of welding current. The movable core, movable coil, and rheostat machines are a little more costly than the simple selector or tap-type machine.

The remote control or electrically adjustable reactor can be either operated by the foot or hand. By means of an extension cable or wire, the remote control can be located next to the operator. The remote control is becoming more popular because it gives the operator more control—especially when welding variable thicknesses of metal. The remote control is especially valuable when it is used with a high frequency

unit for aluminum (which will be mentioned again in the TIG section).

## DUTY CYCLE

All welding machines are rated with a duty cycle. A duty cycle means that a machine can be operated for a certain percent of a given time. For example, if a welding machine is rated at 40% duty cycle, that means it can be operated constantly for 4 out of 10 minutes or 24 minutes out of an hour. If a duty cycle is exceeded, the transformer could overheat and cause damage. One sign of overheating is when the welder, set at 90 amps, gradually seems to lose power. The duty cycle is seldom exceeded because a good bit of time is used to change the electrode, chip the slag, change positions, or observe the welds. The better machines are rated higher and some are rated at 100% at certain amperages.

The majority of the lower priced welding machines use aluminum instead of copper for the transformer winding. This came about basically for two reasons—aluminum is cheaper and a good conductor; and some time ago, due to a strike, copper became scarce, causing the

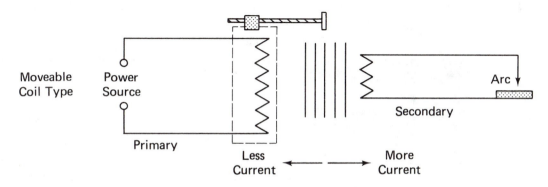

**Figure 8-9** Another AC welding machine.

manufacturers to substitute aluminum for copper.

Aluminum winding will heat up faster than copper winding during the same period of time with the same amperage. In order to protect the transformer, the duty cycle was decreased in the machines using aluminum. After a number of years of service, some machines must be set at a higher amperage to get the same heat that was produced at a lower amperage when the machine was new. The majority of the machines used for constant duty welding still use copper for the transformer windings.

## DC MACHINES

The first welding machines produced were DC machines. Many of the earlier models were generators driven by an electric motor contained in the same cabinet or case (see Figure 8-10). DC current means the current flow is one

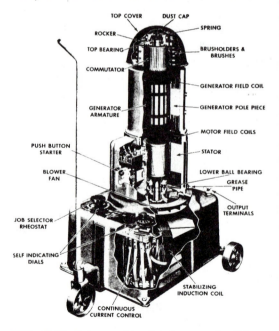

TOP COVER
DUST CAP
ROCKER
SPRING
TOP BEARING
BRUSHOLDERS &
BRUSHES
COMMUTATOR
GENERATOR FIELD COIL
GENERATOR
ARMATURE
GENERATOR POLE PIECE
MOTOR FIELD COILS
PUSH BUTTON
STARTER
STATOR
BLOWER
FAN
LOWER BALL BEARING
GREASE
PIPE
OUTPUT
TERMINALS
JOB SELECTOR
RHEOSTAT
SELF INDICATING
DIALS
STABILIZING
INDUCTION COIL
CONTINUOUS
CURRENT CONTROL

**Figure 8-10** Cutaway view of motor-generator DC arc welder. (*Courtesy of Lincoln Electric.*)

direction only. The current can be made to travel in either direction. In discussing DC machines, DCSP means direct current, straight polarity. The electrons in straight polarity are flowing from the negative terminal (see Figure 8-11) of the welder (cathode) to the electrode. The electrons continue to travel across the arc into the base metal, which is the positive terminal (anode) of the welder. Approximately 2/3 of the total heat produced with DCSP is released at the base metal with 1/3 being released at the electrode.

In some applications, this is the most desirable current because of the heat factor. The DC welder generally produces a softer arc with less penetration, as in welding thin sheet metal and high speed welding, because of the rapid burn off rate of the electrode.

DCRP means direct current, reversed polarity. It is sometimes desirable to reverse the flow of electrons in the welding circuit. When electrons flow from the negative terminal (cathode) of the welder to the base metal (anode), this circuit is known as reversed polarity (see Figure 8-12). In this case, the electrons return to the positive terminal (anode) of the machine from the electrode side of the arc.

When using DCRP, 1/3 of the heat is generated in the arc and released at the base metal and 2/3 of the heat is liberated at the electrode. This gives a more forceful digging arc with deep penetration. The electrode metal and the shielding gas are superheated. This causes the electrode metal to cross the air gap at a great rate of speed. One theory is that the jet action of the gases propels the metal across the arc with great impact.

### DISADVANTAGES—ARC BLOW

One of the disadvantages of the DC welder is a phenomenal condition called ARC BLOW. When a current, such as the welding current, passes through the welding area of the work

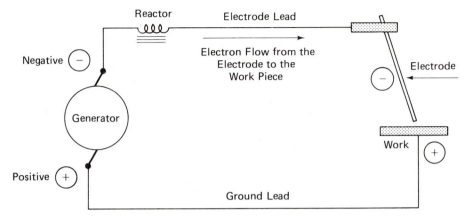

**Figure 8-11** DC welding circuit. Direct current, straight polarity (DCSP) circuit.

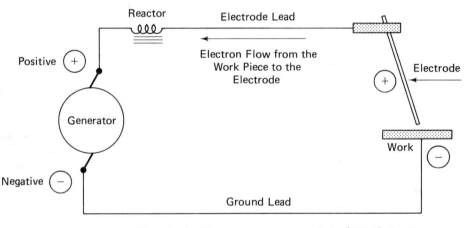

**Figure 8-12** DC welding circuit. Direct current, reverse polarity (DCRP) circuit.

piece, a magnetic line of force builds up to form a magnetic field in the work piece. When these lines of force become concentrated at the end of the weld or bead or as the electrode nears a corner, they cause the arc to wander or deflect from the normal path of travel (see Figure 8-13). Several ways of reducing arc blow are by changing the location of the ground cable, placing a piece of metal against the edge of the metal being welded, changing direction of travel, using two grounds (one before and one after), or changing the angle of the electrode.

## TYPES OF DC MACHINES

There are two basic types of DC machines. The motor driven or electric driven, and the rectifier type. The most commonly used one for average repairing is the rectifier type.

For the most part, the DC machine is very similar to the AC transformer machine (see

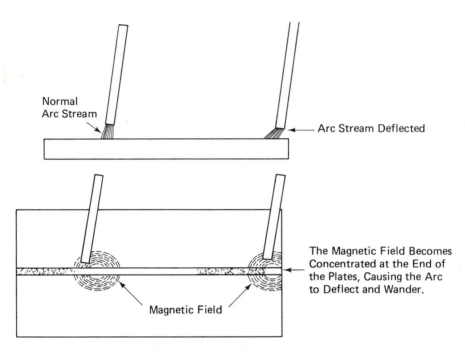

**Figure 8-13** Side and top views. DC welding magnetic arc blow.

Figure 8-14), but the DC machine uses a rectifier to convert AC to DC. Silicon or selenium rectifiers are generally used. They are semiconductors of electricity; they permit easy flow of current in one direction but very little in the opposite direction. These can be compared to the alternator on an automobile which produces AC current and is rectified by diodes to DC output. The silicon rectifier (about 3/4" in diameter) is used for current up to 275 amps (see Figure 8-15). There are no moving parts except for the cooling fan used to cool the transformer. The changing of the polarity is made by a switch or by plug-in cable ends (see Figure 8-16).

The motor driven types are generally much larger in design, and they generate their own current (similar to Figure 8-10). They provide coarse and fine amperage control and continuous duty type, but they cost much more and are noisier than the rectifier type. The portable welders with gasoline engines are very

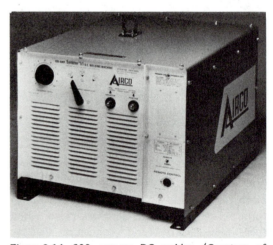

**Figure 8-14** 300 ampere DC welder. (*Courtesy of Airco.*)

popular, but these are available in AC or ACDC (see Figure 8-17).

Many models of welders are made to produce either AC or DC by means of a simple switch; some machines have AC or DC plug-in

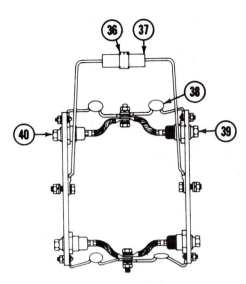

**Figure 8-15** Rectifier assembly for a 150 amp AC-DC welder. 36: Clamp, mtg-capacitor; 37: Capacitor, 0.5 mfd 200 volt DC; 38: Capacitor; 39: Diode, 66-9390 straight polarity; 40: Diode, 66-9391 reverse polarity. (*Courtesy of Airco.*)

type cable taps. Combined with some models is a high frequency circuit built in to operate the TIG torch for aluminum welding.

## WELDING CABLES, POWER CABLES AND OTHER EQUIPMENT

### *POWER CABLES*

Power cables carry the current from the wall receptacle or switch box to the welding machine. Many of the welding machines that are sold come with the power cables—sometimes called the "whip." In this case, the size of the wire in the cable should be adequate to supply the welding machine with sufficient current when operating at its maximum. It is possible that a manufacturer might install an undersize wire cable, which would cut down on the efficiency of the welding machine.

When you purchase welders without power cables, you should install the size of cable recommended by the equipment manufacturer.

If the power source is located some distance from the welding machine, remember that the size of the cable is increased as the distance gets greater. This also applies to extension cables (Figure 8-18) that are often made to be used when the welding job cannot be moved nearer to the welding machine. One size larger is better than one size smaller.

### *WELDING CABLES*

Like power cables, many of the welding machines come with a welding cable already attached to the machine. Other machines come with the welding cable not attached, which means attaching the cables to the machine with a bolt-on lug or a tap connection. Generally, these cables should be adequate for welding at the maximum of the machine. If the cables need replacing, order the same size as the original cable. Undersized cables will overheat and cut down the current to the weld area.

### *ELECTRODE HOLDERS*

Electrode holders with the proper rated amperage are generally supplied with many of the welding machines. Very often, these holders wear out or become worn and should be replaced. The new ones should be rated at the same amperage, or should exceed the amperage, of the machine. Many experienced welders prefer to purchase a certain type most suited to them or their hand, which they feel comfortable with—especially if the holder is being used for long periods of time. There are many different types of holders available in amperage rating, electrode maximum size, and styles that are well insulated.

For beginners, the spring type (see Figure 8-19) is best because the electrode can be released by squeezing the handle in case the electrode becomes stuck to the metal. The twist tight type (see Figure 8-20), of electrode holder might cause a problem because the head must be turned to release the electrode. Some

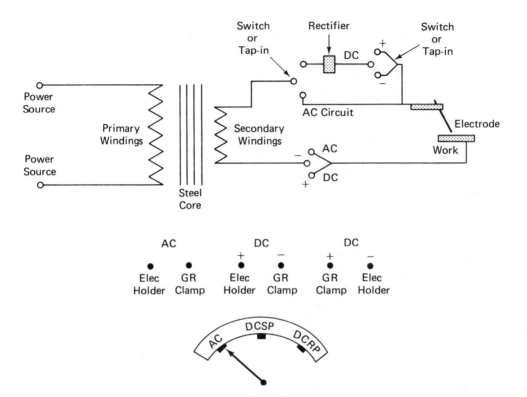

**Figure 8-16** Basic circuit for AC-DC welder. Many welders can be switched from AC to DC or DC to AC by using tap-ins. For switching polarity in DC, change electrode lead from + to − or − to +. Some welders use a master control lever or dial when switching from AC to DCSP or DCRP. Some different types of AC-DC welders.

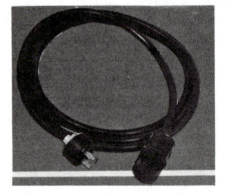

**Figure 8-18** Extension cable. (*Courtesy of Marquette Division of Applied Power.*)

**Figure 8-17** Portable gasoline powered AC-DC welder, 210 amps output with 115 AC volts outlets for operating power tools. (*Courtesy of Lincoln Electric.*)

professional welders prefer this type when working in tight or awkward places because the electrode cannot slip out of the holder.

Figure 8-19 Spring type electrode holder. (*Courtesy of Hobart Brothers.*)

Figure 8-20 Twist tight electrode holder. (*Courtesy of Hobart Brothers.*)

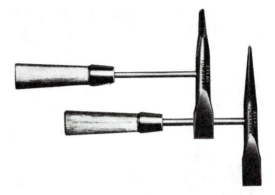

Figure 8-22 Chipping hammer with chisel and cross chisel or point. (*Courtesy of Hobart Brothers.*)

Figure 8-21 Use a good ground clamp. (*Courtesy of Hobart Brothers.*)

Figure 8-23 Chipping hammer with wire brush. (*Courtesy of Hobart Brothers.*)

When the spring tension does not hold the electrode tight or the insulation becomes cracked or broken, replace the holder with a new one. This could save time and possible injury.

## GROUND CLAMPS

Ground clamps (see Figure 8-21) are very important to the welding circuit, not only in electric arc but in MIG and TIG welding. Many of the problems with welding machines or the inability to make proper welds are sometimes caused by the ground connection. This could be the ground clamp improperly attached to the work piece or a poor connection between the cable and the ground clamp. One indication of a poor cable connection is when the area by the cable connection heats up after welding has

Figure 8-24 Wire brushes available in steel and stainless steel. (*Courtesy of Hobart Brothers.*)

been going on for awhile.

Through normal use, the ground clamps lose their spring tension or the teeth on the contact surface wear down reducing the gripping power of the clamp. If that is the case, install a new one. When all of the welding is performed in a particular spot such as a steel workbench or platform, it is better to remove the clamp and

**Figure 8-25** Needle scaler—air operated, used for removing rust, paint, and dirt. (*Courtesy of Hobart Brothers.*)

**Figure 8-27** Wire cup brushes. (*Courtesy of Black & Decker.*)

**Figure 8-26** Electric heavy duty sander/grinder. Caution: Wear safety glasses and use recommended guards. (*Courtesy of Black & Decker.*)

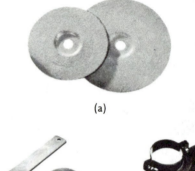

(a)

(b)                                              (c)

**Figure 8-28** (a) Depressed center wheels; (b) disc wheel adapter; (c) guard. (*Courtesy of Black & Decker.*)

attach the ground lead to the bench or platform securely. If the cable has lug end, either bolt it to the bench or weld a bolt to the bench in an out-of-the-way spot and attach the ground lead.

There are special clamps made for that purpose, or a good C-clamp or vise grip can be used to securely fasten the ground lead. Remember, a good welding circuit depends on good connections.

## OTHER EQUIPMENT

**Chipping Hammers with or without Wire Brush** Chipping hammers are used to remove the slag after welding or before another bead can be welded over the first one. Attempting to weld over slag could cause porosity and/or gas bubbles in the weld metal. The chipping hammer with or without brushes has a chisel and a cross chisel or a pointed end which is ideal for removing slag in the deep holes in the weld metal (see Figures 8-22 and 8-23). After using a chipping hammer, a good wire brush must be used to remove any traces of slag.

The wire brush and chipping hammer should be used to remove rust, scale, paint, or any other foreign material before welding to avoid this material being trapped in the weld metal. A stainless steel brush (Figure 8-24) should be used when welding stainless steel, aluminum and other metals. The air-operated needle scaler (Figure 8-25) does a quick job on paint, rust, and other materials as well as slag. It is excellent for digging out the crevices.

(a)                    (b)

Figure 8-29 (a) Flaring cup wheel; (b) wheel guard. (*Courtesy of Black & Decker.*)

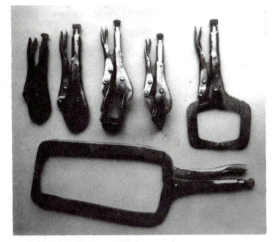

Figure 8-31 Assortment of vise grips. (*Courtesy of Hobart Brothers.*)

Figure 8-30 General service C-clamps. (*Courtesy of Proto Tool Co.*)

Figure 8-32 Pony steel bar clamp. (*Courtesy of Hobart Brothers.*)

**Power Grinders** Power grinders, either air or electric (see Figure 8-26), operated with wire cup brushes (Figure 8-27) for removing rust and foreign material, are good to use. Grinding stones in the shape of depress (Figure 8-28) center and cup wheels (Figure 8-29) can be used for beveling plate, trimming metal edges, or finishing off welds or beads.

**Optional Equipment** Other equipment, although optional but very useful, includes several light and heavy hammers for bending. An assortment of C-clamps (Figure 8-30) and vise grips (Figure 8-31) used for holding metal together or in position can be helpful. When working on larger objects, a set of bar clamps (Figure 8-32) is very useful.

## QUESTIONS

1. What makes the AC transformer type welder very popular?
2. What has led to the increased popularity of the AC welder?
3. What type transformers are generally used in arc welding?

4. What is open circuit potential?

5. What are some advantages of the AC welder?

6. Name some disadvantages of AC welders?

7. While changing direction in the AC current welding circuit, what helps to maintain the arc when the voltage drops to 0?

8. In theory, what percentage of heat is released at the electrode?

9. In the AC transformers, which windings are heavier than the others?

10. What type coil is used to add to arc stability?

11. What types of AC machines have a limited selection of current settings?

12. What is the advantage of the movable coil and core welding machines?

13. What is the main advantage of the remote control or electrically adjustable reactor?

14. What is a duty cycle?

15. What two metals are used for transformer winding?

16. What were the first types of DC welding machines used?

17. What does DCSP, or direct current, straight polarity, mean?

18. What is DCRP?

19. Which one produces the most heat at the electrode?

20. Which one produces a softer arc?

21. What is one disadvantage of the DC welder?

22. What are the two basic types of DC Machines?

23. How do the rectifiers change AC current to DC current?

24. Do undersize power cables have any effect on welding machine efficiency?

25. Should replacement cables or electrode holders exceed the welding machine maximum output?

26. Can a defective ground clamp or connection cause welding problems?

27. What can be used to remove welding slag?

# ELECTRODES

## chapter 9

The electrode provides the filler metal for the weld (see Figure 9-1). The first electrode produced was the bare electrode. Bare electrodes are the least expensive and are still being used for some welding operations. However, it is more difficult to produce a satisfactory weld with a bare electrode. A good welder can produce a satisfactory weld with them, but the strength and durability of the joint is less than that produced by the coated electrode.

The *dust cover electrode* uses the coating to help stabilize the arc. Some are coated with a light coating of lime. These rods produce little or no slag on the weld and reduce the oxidizing action of the arc.

*Flux dipped electrodes* are bare wires dipped into liquid flux, which is baked on. The flux is often uneven and occasionally there is too much flux on the rod. Some aluminum welding electrodes are of the dipped type.

The most common type used is the *extruded type coated electrode*. The flux coating is put on by forcing the welding electrode wire through a die and then baking the coating.

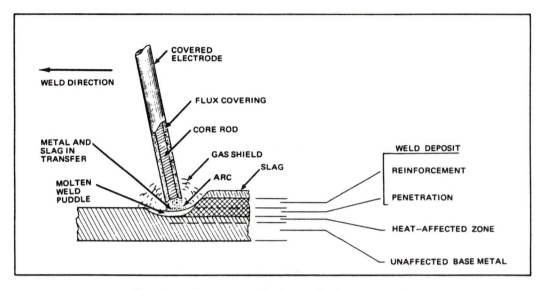

**Figure 9-1** Electrode deposit thru arc. (*Courtesy of Airco.*)

Most of the mild steel electrodes are physically similar in their specifications except for the flux being used. These electrodes range from 1/16" to 3/8" (1.6 mm to 9.52 mm) in diameter—not including the coating.

## FUNDAMENTALS OF THE ELECTRODE COATING

The purpose of the coating on the electrode is to provide a gaseous shield around the arc during the welding process to expel any oxygen, nitrogen, or other impurities that might combine with the molten metal (see Figure 9-1). The coating also acts as cleaning agent to prevent impurities from remaining in the molten puddle of metal, which could cause weakness of the weld.

The coating also provides a coating of slag resulting from the oxidation of the materials. The oxidized slag solidifies on top of the weld material and protects it until the metal cools.

This prevents other impurities from affecting the welded metal.

Some of the materials contained in the coating are silicon, calcium in the form of calcium carbonate, magnesium carbonate, asbestos, titanium dioxide cellulose, metallic oxides, and so on. The principal stabilizers used in most coatings are feldspar, titanium, and calcium carbonate.

NOTE: Electrodes must be stored in a dry place to prevent moisture from destroying the coating. Also they must be handled carefully to prevent the coating from being cracked or chipped off.

## CLASSIFICATION OF ELECTRODES

Until the start of World War II, there were no set standards for the manufacture of electrodes. At that time, the American Welding Society in conjunction with the American Society for Testing Materials, developed a standard governing the manufacture of electrodes for mild steel and later one for alloy and carbon steel and other ferrous and non-ferrous metals. The user now has information on which electrode can produce a welded metal of certain mechanical property, the position of the weld, penetration, and the type of current used.

All electrodes are given a number containing four or five digits preceded by the letter E, which signifies electrode. If an electrode is marked E6011 or E10016, the numbers can be broken down as follows: E is for electrode; 60 or 100, the first two or three digits mean the minimum tensile strength in thousands pounds per square inch. This is the resistance of a material to forces trying to pull it apart. Most rods will exceed the minimum strength. For example, 60 means 60,000 psi (414,000 kPa) tensile strength. The third or fourth digit is the position possible for welding:

1—all positions—flat, vertical, horizontal, or overhead

2—horizontal and flat positions only

3—flat position only

The fourth (or fifth) digit is the type of coating and the welding current:

| | |
|---|---|
| 0—cellulose sodium | DCRP only |
| 1—cellulose potassium | AC or DCRP |
| 2—titania sodium | AC or DCSP |
| 3—titania potassium | AC or DCSP and DCRP |
| 4—iron powder titania | AC or DCSP and DCRP |
| 5—low hydrogen sodium | DCRP only |
| 6—low hydrogen potassium | AC or DCRP |
| 7—iron powder, iron oxide | AC or DCSP or DCRP |
| 8—iron powder, low hydrogen | AC or DCRP |

## WELDING CHARACTERISTICS OF THE ELECTRODES

0—deep penetration, flat bead, thin slag, good in all positions

1—deep penetration, flat bead, thin slag, good in all positions

2—medium penetration, heavy slag, convex bead

3—shallow penetration, medium slag, convex bead, general purpose

4—medium penetration, thick slag, convex bead, fast deposit rate

5—medium penetration, medium slag, convex bead

6—medium penetration, medium to heavy slag, convex bead, high carbon steel

7—medium penetration, flat beads

8—shallow penetration, convex bead

## IRON POWDER ELECTRODES

Iron powder has been added to the coating of many basic electrodes. The iron powder is converted to steel by the heat of the arc, thus adding additional metal to the weld deposit. These electrodes are designed for high speed welding; the bead appearance is improved. The heavy slag is easily removed.

## LOW-HYDROGEN ELECTRODES

Hydrogen has a critical effect on the weld deposits involving welding high tensile carbon steels. If the hydrogen is trapped in the welded metal, under-bead cracking could result. These low-hydrogen coated electrodes are made to prevent the introduction of hydrogen into the weld.

| Coating Color | Conforms to Test Requirements of AWS Class | Electrode Brand Name | Electrode Polarity (+)="REVERSE" (−)="STRAIGHT" | 5/64" Size | 3/32" Size | 1/8" Size | 5/32" Size | 3/16" Size | 7/32" Size | 1/4" Size | 5/16" Size |
|---|---|---|---|---|---|---|---|---|---|---|---|
| **MILD STEEL** | | | | | | | | | | | |
| Light Tan | E6010 | Fleetweld 5 | DC (+) | | | | | | 200-275 | 250-325 | 280-400 |
| Brick Red | E6010 | Fleetweld 5P | DC (+) | | 40-75 | 75-130 | 90-175 | 140-225 | | | |
| Tan | E6012 | Fleetweld 7 | DC (−) / AC | | | 80-135 / 90-150 | 110-180 / 120-200 | 155-250 / 170-275 | 225-290 / 250-320 | 245-325 / 275-360 | |
| Light Tan | E6011 | Fleetweld 35 | AC / DC (+) | | | 75-120 / 70-110 | 90-160 / 80-145 | 120-200 / 110-180 | 150-260 / 135-235 | 180-300 / 170-270 | |
| Red Brown | E6011 | Fleetweld 35LS | AC / DC (±) | | | 80-130 / 70-120 | 120-160 / 110-150 | | | | |
| Dark Tan | E6013 | Fleetweld 37 | AC / DC (±) | | 75-105 / 70-95 | 100-150 / 90-135 | 150-200 / 135-180 | 200-260 / 180-235 | | | |
| Gray-Brown | E7014 | Fleetweld 47 | AC / DC (−) | | | 110-160 / 100-145 | 150-225 / 135-200 | 200-280 / 180-250 | 260-340 / 235-305 | 280-425 / 260-380 | |
| Brown* | E6013 | Fleetweld 57 | AC / DC (±) | 45-80 / 40-75 | 75-105 / 70-95 | 100-150 / 90-135 | 150-200 / 135-180 | 200-260 / 180-235 | 250-310 / 225-280 | 300-360 / 270-330 | 360-460 / 330-430 |
| Brown | E6011 | Fleetweld 180 | AC / DC (+) | | 40-90 / 40-80 | 60-120 / 55-110 | 115-150 / 105-135 | | | | |
| Gray | E7024 | Jetweld 1 | AC / DC (±) | | 65-120 / 60-110 | 115-175 / 100-160 | 180-240 / 160-215 | 240-300 / 220-280 | 300-380 / 270-340 | 350-440 / 320-400 | |
| Brown | E6027 | Jetweld 2 | AC / DC (±) | | | | 190-240 / 175-215 | 250-300 / 230-270 | 300-380 / 270-340 | 350-450 / 315-405 | |
| Gray* | E7024 | Jetweld 3 | AC / DC (±) | | | 115-175 / 100-160 | 180-240 / 160-215 | 240-315 / 215-285 | 300-380 / 270-340 | 350-450 / 315-405 | 450-600 |
| Gray | E7018 | Jetweld LH-70 | DC (+) / AC | | 70-100 / 80-130 | 90-150 / 110-170 | 120-190 / 135-225 | 170-280 / 200-300 | 210-330 / 260-380 | 290-430 / 325-440 | 375-500 / 400-530 |
| Gray* | E7018 | Jet-LH 72 | DC (+) / AC | | 70-100 / 80-120 | 85-150 / 100-170 | 120-190 / 135-225 | 190-260 / 210-300 | 250-330 / 270-370 | 300-400 / 325-420 | |
| Gray | 7018 (white numbers) | JET-LH 78 | DC (+) / AC | | 80-100 | 100-150 / 110-170 | 130-190 / 140-225 | 180-270 / 210-290 | 250-330 / 270-370 | 300-400 / 325-420 | |
| Gray-Brown | E7028 | Jetweld LH-3800 | AC / DC (+) | | | | 180-270 / 170-240 | 240-330 / 210-300 | 275-410 / 260-380 | 360-520 | |
| **LOW ALLOY, HIGH TENSILE STEEL** | | | | | | | | | | | |
| Pink | E7010-A1 | Shield-Arc 85 | DC (+) | | 50-90 | 75-130 | 90-175 | 140-225 | | | |
| Pink* | E7010-A1 | Shield-Arc 85P | DC (+) | | | | | 140-225 | | | |
| Tan | E7010-G | Shield-Arc HYP | DC (+) | | | 75-130 | 90-185 | 140-225 | 160-250 | | |
| Tan* | E7010-G | Shield-Arc 65+ | DC (+) | | | 75-130 | 90-185 | 140-225 | | | |
| Tan | E8010-G | Shield-Arc 70+ | DC (+) | | | 75-130 | 90-185 | 140-225 | | | |
| Gray-Brown | E8018-C1 | Jet-LH 8018-C1 | DC (+) / AC | | | 90-150 / 110-160 | 120-180 / 140-200 | 180-270 / 200-300 | | 250-350 / 300-400 | |
| Gray-Brown | E8018-C3 | Jet-LH 8018-C3 | DC (+) / AC | | | 90-150 / 110-160 | 120-180 / 140-200 | 180-270 / 200-300 | 210-330 / 250-360 | 250-350 / 300-400 | |
| Gray | E8018-B2 | Jetweld LH-90 | DC (+) / AC | | | 90-150 / 110-160 | 110-180 / 140-230 | 160-280 / 200-310 | | | |
| Gray | E11018-M | Jetweld LH-110M | DC (+) / AC | | | 95-155 / 115-165 | 120-180 / 145-230 | 160-280 / 200-310 | 190-310 / 240-350 | 230-360 / 290-410 | |
| **STAINLESS STEEL** | | | | | | | | | | | |
| Pale Green | E308-15 | Stainweld 308-15 | DC (+) | | 30-70 | 50-100 | 75-130 | 95-165 | | 150-225 | |
| Gray | E308-16 | Stainweld 308-16 | DC (+); AC | 20-45 | 30-60 | 55-95 | 80-135 | 115-185 | | 200-275 | |
| Gray | E308L-16 | Stainweld 308L-16 | DC (+); AC | | 30-65 | 55-100 | 80-140 | 115-190 | | | |
| Gray | E309-16 | Stainweld 309-16 | DC (+); AC | | 30-60 | 55-95 | 80-135 | 115-185 | | 200-275 | |
| Pale Green | E310-15 | Stainweld 310-15 | DC (+) | | 30-70 | 45-95 | 80-135 | 100-165 | | | |
| Gray | E310-16 | Stainweld 310-16 | DC (+); AC | | 30-65 | 55-100 | 80-140 | 120-185 | | 200-275 | |
| Gray | E316L-16 | Stainweld 316L-16 | DC (+); AC | | 30-65 | 55-100 | 80-140 | 115-190 | | | |
| Pale Green | E347-15 | Stainweld 347-15 | DC (+) | | 30-70 | 50-100 | 75-130 | 95-165 | | | |
| Gray | E347-16 | Stainweld 347-16 | DC (+); AC | | 30-60 | 55-95 | 80-135 | 115-185 | | | |
| **BRONZE & ALUMINUM** | | | | | | | | | | | |
| Peach | E-CuSn-C | Aerisweld | DC (+) | | | | 50-125 | 70-170 | | 90-220 | |
| White | Al-43 | Aluminweld | DC (+) | | | 20-55 | 45-125 | 60-170 | | 85-235 | |

| Coating Color | Conforms to Test Requirements of AWS Class | Electrode Brand Name | Electrode Polarity | 1/8" Size | 5/32" Size | 3/16" Size | 1/4" Size |
|---|---|---|---|---|---|---|---|
| **CAST IRON** | | | | | | | |
| Light Tan | ESt | Ferroweld | DC (+); AC | 80-100 | | | |
| Black | ENi-Cl | Softweld | DC (±) / AC | 60-110 / 65-120 | 100-135 / 110-150 | | |

| Coating Color | Conforms to Test Requirements of AWS Class | Electrode Brand Name | Electrode Polarity | 5/64" Size | 3/32" Size | 1/8" Size | 5/32" Size | 3/16" Size | 1/4" Size |
|---|---|---|---|---|---|---|---|---|---|
| **HARDSURFACING** | | | | | | | | | |
| Black | | Abrasoweld | DC (±) / AC | 40-150 / 50-165 | 75-200 / 80-220 | 110-250 / 120-275 | 150-375 / 165-410 | | |
| Black | | Faceweld 1 | DC (+); AC | | | 60-150 | | | |
| Black | | Faceweld 12 | DC (+); AC | | | 60-150 | | | |
| Dark Gray | | Jet-LH BU-90 | DC (±) / AC | | | 145-210 / 155-225 | 180-280 / 200-290 | 230-360 / 255-375 | |
| Dark Gray | | Mangjet | DC (±) / AC | | | 120-180 / 135-230 | 160-260 / 165-285 | 200-350 / 220-385 | |
| Dark Gray | | Wearweld | DC (+) / AC | | | | 110-275 / 125-275 | 150-400 / 200-400 | |

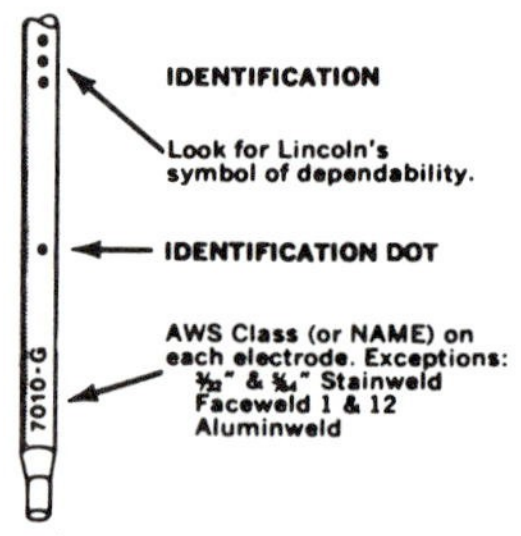

**IDENTIFICATION**

Look for Lincoln's symbol of dependability.

**IDENTIFICATION DOT**

AWS Class (or NAME) on each electrode. Exceptions: 3/32" & 5/64" Stainweld Faceweld 1 & 12 Aluminweld

7010-G

®=Means registered Trademark of The Lincoln Electric Company
*Has identification dot on coating

**Figure 9-2**  Electrode identification chart. (*Courtesy of Lincoln Electric.*)

There are many different types of electrodes available for welding cast iron, stainless steels, aluminum, and other metals and alloys. Also, there are electrodes made for hard surfacing, impact resistance and many low temperature electrodes used where normal welding heat range electrode could damage the metal grain structure.

Occasionally, some electrodes will have a suffix following the last digit, for example E8018-C1 or 10016B3. This means that the normal alloying agent has been altered. For example, C1 has 2.5% nickel, while B3 has 2.25% chromium plus 1% molybdenum.

## ELECTRODE IDENTIFICATION

Electrodes are identified by several means. Some have painted spots at the end of the electrode called the grip end (see Figure 9-2). Some are painted on the grip area while others are painted near the grip end on the coating. Some manufacturers print the AWS number on the coating, although some use their own trade name or number on the coating, or they have their own coating color system (see Figure 9-3). For best results, consult the manufacturer's handbook. Many electrodes lying in a small shop are difficult to identify: what they are and what they are used for.

See chart (Figure 9-4) for comparative index of mild steel and low-hydrogen–low alloy electrodes.

*Mild steel electrodes* may be classified into the following groups:

FAST-FREEZE[1] GROUP (6010, 6011)— Electrodes which have a deep penetrating arc and fast freeze. Fast freeze means that the molten puddle will solidify very rapidly.

FILL-FREEZE[1] GROUP (6012, 6013, 6014, 7014)—Electrodes with a moderate forceful

[1] Lincoln welding.

## TYPICAL MECHANICAL PROPERTIES

Test procedures per appropriate AWS spec. Other specs may have different min. requirements and test procedures. ③
Low figures in the as welded ranges below are AWS minimum requirements.

|  | FLEETWELD 7 | FLEETWELD 37 | FLEETWELD 57 | FLEETWELD 47 |
|---|---|---|---|---|
| AS WELDED<br>Tensile Strength—psi | 67-79,000 | 67-74,000 | 67-77,000 | 72-76,000 |
| Yield Point—psi<br>Ductility—% of Elong. in 2" | 55-70,000<br>17-22 | 55-68,000<br>17-28 | 55-71,000<br>17-26 | 60-70,000<br>17-28 |
| Charpy V-Notch Toughness—Ft. Lbs. | 58 @ 70°F | 70 @ 70°F | 55 @ 70°F | 55 @ 70°F |
| STRESS RELIEVED @ 1150°F<br>Tensile Strength—psi | 67-84,000 | 67-74,000 | 67-70,000 | 67-77,000 |
| Yield Point—psi<br>Ductility—% of Elong. in 2" | 55-73,000<br>17-24 | 55-68,000<br>17-29 | 60-61,000<br>24-30 | 55-70,000<br>24-30 |
| Charpy V-Notch Toughness—Ft. Lbs. | 62 @ 70°F |  | 55 @ 70°F · | 48 @ 70°F |

③ Other tests and other procedures may produce different results (see "Properties vs. Procedures" — page 2.)

## CONFORMANCES AND APPROVALS

See Lincoln Price Book for certificate numbers, size and position limitations and other data.

| Conforms to Test Requirements of<br>AWS—A5.1 and ASME—SFA5.1 | E6012 | E6013 | E6013 | E7014 |
|---|---|---|---|---|
| ASME Boiler Code { Group<br>{ Analysis | F2<br>A1 | F2<br>A1 | F2<br>A1 | F2<br>A1 |
| American Bureau of Shipping<br>& U. S. Coast Guard | Approved | Approved |  | Approved |
| Conformance Certificate Available ④ | Yes | Yes | Yes | Yes |
| Lloyds | Approved | Approved | Approved | Approved |
| Military Specification | MIL-QQE-450 | MIL-QQE-450 |  |  |

④ "Certificate of Conformance" to AWS classification test requirements is available. These are needed for Federal Highway Administration (formerly Bureau of Public Roads) projects.

Figure 9-3  Important information to check when selecting electrodes. (*Courtesy of Lincoln Electric.*)

| AWS CLASS | HOBART | AIR PRODUCTS | AIRCO | ARCOS | ALLOY RODS | LINCOLN | MARQUETTE | McKAY CO. | REID-AVERY CO. (RACO) | UNIBRAZE | WESTINGHOUSE |
|---|---|---|---|---|---|---|---|---|---|---|---|
| E-6010 | 10 / 60AP | AP-6010-IP | 6010 | — | SW-610 | Fleetweld 5 / Fleetweld 5P | — | 6010 / 60101P | 6010 | E6010 | XL-610 / ZIP 10 |
| E-6011 | 335A | AP-6011 / 6011-C | 6011 LOC / 6011 / 6011C | — | SW-14 / SW-14 IMP / Steel Arc | Fleetweld 35, 180 / 35LS | 130 | 6011 / 6011-IP | 6011 / 6011-IP | E6011 | ACP-611 |
| E-6012 | 12 / 212A / 12-A | AP-6012-GP / 6012-SF / 6012-IP | 6012 / 6012C | — | SW-612 / PFA | Fleetweld 7 | — | 6012 | 6012 / 6012-F | E6012 | FP-612 / FP-2-612 |
| E-6013 | 413 / 447-A | AP-6013-GP / 6013-SF | 6013 / 6013C | — | SW-16 / SW-15 / Steel Arc Plus | Fleetweld 37, 57 | 140 | 6013 | 6013 | E6013 | SW-613 / SW-2M-613 |
| E-7014 | 14A | AP-7014-IP | Easyarc 7014 | — | SW-15-IP / Monoweld | Fleetweld 47 | 146 | 7014 | 7014 | E7014 | ZIP-14 |
| E-7024 | 24 | AP-7024-IP | Easyarc 7024 | — | SW-44 | Jetweld 1 / 3 | 24 | 7024 | — | E7024 | ZIP-24 |
| E-7028 | 728 | — | Easyarc 7028 | — | — | Jetweld LH-3800 | — | — | Raco 7028 | E7028 | — |
| E-7018 | 718 / 718-SR / 718LMP | AP-7018 / 7018IP | Easyarc 7018, 7018C, 7018-A1, Code Arc 7018 | Ductilend 70 | Atom-Arc 7018, 170-LA, SW-47, 616 | Jetweld LH-70 | 7018 | 7018 | 7018 | 7018 | Wiz-18 |
| E-7010-G | 70-AP | — | — | — | — | Shield-Arc 85-P | — | — | Raco 7010-G | — | 7011-A1 |
| E-7018-A1 | LH718-MO | AP-7018-A1 | Code Arc 7018-A1 | Ductilend 70-MO | Atom-Arc 7018-MO / 718-A1 | — | — | 7018-A1 | 7018-A1 | 7018-A1 | 7018-A1 |
| E-8010-G | 80AP | — | — | — | — | 70T | — | — | — | — | — |
| E-8018-C1 | LH-818-NI | AP-8018-C1 | Code Arc 8018 / C1MR | Nickend 2 | Atom-Arc 8018-C1 | Jet LH-8018-C1 | — | 8018-C1 | Raco 8018-C1 | — | 8018-C1 |
| E-8018-C2 | LH-818-N2 | AP-8018-C2 | Code Arc 8018-C2 | | Atom-Arc 8018N | — | — | 8018-C2 | 8018-C2 | — | 8018-C2 |
| E-8018-C3 | LH-818-N3 | AP-8018-C3 | Code Arc 8018-C3 | Ductilend 80 | Atom-Arc 8018, 108 | LH-8018-C3 | — | 8018-C3 | Raco 8018-C3 | — | 8018-C3 |
| E-8018-B2 | LH-818-CM | AP-8018-B2 | Code Arc 8018-B2 | Chromend 1-MA | Atom-Arc 8018CM / 180-LE | LH-90 | — | 8018-B2 | Raco 8018-B2 | — | 8018-B2 |
| E-9010-G | 90AP | — | — | — | — | — | — | — | — | — | — |
| E-9018-M | LH-918-M | AP-9018-G | — | Ductilend 85 | Atom-Arc 9018, 109 | | — | 9018 | Raco 9018-M | — | 9018-M |
| E-9018-B3 | LH918-CM | AP-9018-B3 | Code Arc 9018-B3 | | Atom-Arc 9018CM / 190-LE | — | — | 9018-B3 | Raco 9018-B3 | — | 9018-B3 |
| E-10018-D2 | LH-1018 | AP10018-D2 | — | — | Atom-Arc 10018MM | — | — | 10018-D2 | 10018-D2 | — | 10018-D2 |
| E-10018-M | LH-1018-M | AP-10018-G | — | Ductilend 100 | Atom-Arc 10018, 112 | — | — | 10018 | — | — | 10018-M |
| E-11018-M | LH-1118 | — | Code Arc 11018 | Ductilend 110 | Atom-Arc T, 107 | Jetweld LH-110 | — | 11018 | 11018-M | — | 11018-M |
| E-12018-M | LH-1218 | AP-12018-G | Code Arc 12018 | Ductilend 120 | 117 | — | — | 12018 | 12018-M | — | 12018-M |

**Figure 9-4** Comparative index of mild steel and low hydrogen–low alloy electrodes. (*Courtesy of Hobart Brothers*.)

arc and a deposit rate between those of the fast-freeze and fast-fill electrodes. Complete slag coverage and distinct even tipples. Fill-freeze means the molten weld solidifies fast and has a higher deposit rate than the fast-freeze group.

FAST-FILL[1] GROUP (6024, 6027, 7024)— This group includes the heavy coated, iron powder electrodes with their soft arc and fast deposit rate. They produce exceptionally smooth bead and a heavy slag, which is easily removed. This group is used for production work and down hand position welding. Fast fill means that they have a high deposit rate of weld metal.

LOW-HYDROGEN GROUP (E6018, 7018, 6028, 7028)—Low-hydrogen electrodes offer these benefits: outstanding crack resistance, elimination of porosity on sulphur-bearing steels, and x-ray quality deposits. The name of the electrode derives from the fact that the coating contains little hydrogen in either moisture or chemical form. These electrodes must be kept dry and must be stored in a heated cabinet.

## STORING ELECTRODES

Electrodes should be stored in a dry place, and if possible, in the original containers or boxes. The containers or boxes should have additional identification tags or labels especially on the open end of the container or box, if they are stacked.

Through careless handling or storage, especially in small shops, it is almost impossible to properly identify the electrodes because the boxes are torn apart, soiled, or the electrodes are covered with dust. Also, the electrodes are often exposed to moisture, which destroys the coating, and if low-hydrogen, could affect the weld.

**Figure 9-5** Portable oven for electrodes. Holds 13 lbs. (5.8 kg) of 18″ (457 mm) electrodes. (*Courtesy of Hobart Brothers.*)

There are several different types of storage cabinets available for electrodes. Some of these cabinets have heating elements to dry off the electrodes (see Figures 9-5 and 9-6). When using low-hydrogen or other types of electrodes, a small portable carrycase holding about a box of electrodes, with a heating element, can be used.

Another method of storing electrodes is the use of an old refrigerator, without the refrigeration unit. With some modification for additional strong shelves, many boxes of electrodes can be stored at one time. A 5- or 10-watt bulb will supply sufficient heat to keep the electrodes dry and away from dirt and dust. Some shops

**Figure 9-6** Shop type oven for electrodes. Holds 350 lbs (158.7 kg) of 18″ (457 mm) electrodes. (*Courtesy of Hobart Brothers.*)

keep the refrigerators locked because other valuable or expensive electrodes or solders are kept in there to prevent loss.

Caution must be used when using refrigerators with larger light bulbs or large cabinets because excessive heat for an extended period of time could cause the coating on some electrodes such as E6010 or 6011 which contain cellulose to develop fine hairline cracks. These are not noticeable until they are used and the arc stream becomes erratic. These electrodes should not be stored above 130°F (54°C) for any length of time.

Some electrodes such as low-hydrogen, and iron-powder, or electrodes like 6012 and 6013 can be reconditioned by placing them in an oven for one hour at a specified heat, if the electrodes were accidentally exposed to water and the coating is not damaged. One manufacturer recommends that electrodes like 6012 or 6013 be heated at a temperature of at least 250°F (121°C) for one hour. Low-hydrogen and others are heated to a higher temperature.

One important point to remember: protect the electrodes from moisture and damage to the coating. Electrodes with chipped or cracked coatings should be discarded. Avoid leaving electrodes lying around on benches where they will become damaged.

## QUESTIONS

1. What were the first electrodes used?

2. What is the most common type of electrode used today?

3. Name some of the purposes for the electrode coating.

4. Name a few of the elements or compounds used in the electrode coating.

5. What two organizations set up a standardized system for governing the manufacture of electrodes?

6. In the AWS electrode classification number such as E7024, what do the following mean?

a. Letter E           b. Numbers 70.           c. Number 2.

d. If the above number is 1, what does it mean?       e. Number 4.

7. If an electrode is 10016D2, what does D2 stand for?

8. What is the purpose of iron powder electrodes?

9. How are most electrodes identified?

10. What is a Fast Freeze electrode?

11. What are some of the benefits of a low hydrogen electrode?

12. What type electrode must be kept absolutely dry?

13. What type electrode cannot be heated over 130°F (54°C)?

# SELECTION OF THE PROPER ELECTRODES

## chapter 10

The following factors must be considered in selecting the correct arc welding electrode for each job. The electrode metal should be equal to or greater than the mechanical strength of the base metal or the metal being welded. For mild steel, any of the electrodes in the E60xx or 70xx are generally satisfactory.

The base metal composition is important: is it mild steel, low alloy, high carbon, cast iron, or others? There are electrodes available for the different types of metals. The electrode metal should match the base metal in composition.

The position of the welds is important because some electrodes are made to be used in all positions, while some are made for flat or horizontal positions. Many times the object being welded cannot be placed in a flat or horizontal position. The 3rd (4th) digit in the AWS number indicates the welding position of the electrode.

The type of joint or joint design and the fit are important as to whether to use a deep penetrating electrode or a soft, shallow penetrating electrode. For tight fit up or an edge

Figure 10-1 Selecting the proper electrode for the job at hand. Thin metal would require a shallow penetrating electrode. (*Courtesy of Lincoln Electric.*)

that is not beveled, use a deep penetrating or digging electrode like the E6011. When welding thin material, the electrode should have a soft, light penetration arc similar to the E6013 electrode (see Figure 10-1).

The thickness and shapes of the base metal must be considered to avoid cracking of the weld metal. The electrode metal must possess a certain amount of ductility such as the electrodes in the low hydrogen group like the Exx15, 16, or 18 (see Figure 10-2).

Figure 10-2 An overhead joint that requires certain properties; a low hydrogen electrode may be best suited. (*Courtesy of Lincoln Electric.*)

Figure 10-3 A down hand or flat joint could be welded with a fast fill electrode. (*Courtesy of Lincoln Electric.*)

The type of welding current that is available and the type of welding machine such as AC transformer, AC–DC machine, or DC generator or a portable welder are important. Is the welding machine capable of handling large electrodes if needed? Some electrodes are designed for direct current only; some are designed for both AC and DC. The E6010 is designed for DCRP; the E6011 can be used on either AC or DC. Check the polarity recommended for electrodes using DC because some work better with straight polarity and some with reversed polarity.

Some welding jobs must meet severe service conditions such as low temperature, impact, shock, or others. In many cases, low-hydrogen electrodes can be used; otherwise, select the electrode that matches the base metal properties but also has superior ductility and resistance to impact.

Job conditions are to be considered as to production efficiency. It is possible to use electrodes with high deposition rates if the welds can be made in an ideal position. Iron-powdered electrodes such as Exx24 and others are high-deposit electrodes. The possibility of using large electrodes to save time and using one pass instead of several, should be considered (see Figure 10-3).

Some welding jobs may involve most of the above or some may involve a few, depending on the situation. For example, when repairing road or farm equipment, the conditions would be different than production welding or fabrication.

## QUESTIONS

1. Should the electrode metal equal the mechanical strength of the base metal?
2. What is the relationship between the electrode and base metal composition?
3. Why is the position of the weld important when selecting electrodes?
4. Is it important to know the current requirements of the electrodes?
5. What about job conditions?

# IDENTIFICATION
# OF METALS

# chapter 11

When welding ferrous metal (metals that contain iron) or non-ferrous metals (metals that do not contain iron, such as copper and aluminum), it is to the welder's benefit that the exact composition of the metal to be welded is known.

When using sheet steel, plates, or working from drawing, the classification number is generally marked on the metal. A uniform steel classification has been adopted by the American Iron and Steel Institute (AISI) and the Society of Automotive Engineers (SAE). The AISI number for a certain piece of steel could be C1030. The letter C or prefix to the number indicates how the steel was made or what process was used. The letter C indicates that the open-hearth process was used. If the letter E is used, it indicates the electric furnace process. The first digit of the number 1030 indicates carbon steel, a 5 would be for chromium steel. The second digit, or number, represents the percent of alloy in the steel, and the last two digits indicate the carbon content in the steel. The 30 would have a carbon range of 0.28 to 0.34.

NOTE: Non-ferrous metals such as aluminum use a similar standardized classification system. Likewise, the number indicates the composition of the aluminum alloy and the temper.

| metal / test | low carbon steel | medium carbon steel | high carbon steel | high sulphur steel |
|---|---|---|---|---|
| appearance | DARK GREY | DARK GREY | DARK GREY | DARK GREY |
| magnetic | STRONGLY MAGNETIC | STRONGLY MAGNETIC | STRONGLY MAGNETIC | STRONGLY MAGNETIC |
| chisel | CONTINUOUS CHIP SMOOTH EDGES CHIPS EASILY | CONTINUOUS CHIP SMOOTH EDGES CHIPS EASILY | HARD TO CHIP CAN BE CONTINUOUS | CONTINUOUS CHIP SMOOTH EDGES CHIPS EASILY |
| fracture | BRIGHT GREY | VERY LIGHT GREY | VERY LIGHT GREY | BRIGHT GREY FINE GRAIN |
| flame | MELTS FAST BECOMES BRIGHT RED BEFORE MELTING | MELTS FAST BECOMES BRIGHT RED BEFORE MELTING | MELTS FAST BECOMES BRIGHT RED BEFORE MELTING | MELTS FAST BECOMES BRIGHT RED BEFORE MELTING |
| Spark* *For best results, use at least 5,000 surface feet per minute on grinding equipment. (Cir. x R.P.M / S.F. per Min.) 12 | Long Yellow Carrier Lines (Approx. .20% carbon or below) | Yellow Lines Sprigs Very Plain Now (Approx. .20% to .45% carbon) | Yellow Lines Bright Burst Very Clear Numerous Star Burst (Approx. .45% carbon and above) | Swelling Carrier Lines Cigar Shape |

| metal / test | manganese steel | stainless steel | cast iron | wrought iron |
|---|---|---|---|---|
| appearance | DULL CAST SURFACE | BRIGHT, SILVERY SMOOTH | DULL GREY EVIDENCE OF SAND MOLD | LIGHT GREY SMOOTH |
| magnetic | NON MAGNETIC | DEPENDS ON EXACT ANALYSIS | STRONGLY MAGNETIC | STRONGLY MAGNETIC |
| chisel | EXTREMELY HARD TO CHISEL | CONTINUOUS CHIP SMOOTH BRIGHT COLOR | SMALL CHIPS ABOUT ⅛ in. NOT EASY TO CHIP, BRITTLE | CONTINUOUS CHIP SMOOTH EDGES SOFT AND EASILY CUT AND CHIPPED |
| fracture | COARSE GRAINED | DEPENDS ON TYPE BRIGHT | BRITTLE | BRIGHT GREY FIBROUS APPEARANCE |
| flame | MELTS FAST BECOMES BRIGHT RED BEFORE MELTING | MELTS FAST BECOMES BRIGHT RED BEFORE MELTING | MELTS SLOWLY BECOMES DULL RED BEFORE MELTING | MELTS FAST BECOMES BRIGHT RED BEFORE MELTING |
| Spark* *For best results, use at least 5,000 surface feet per minute on grinding equipment. (Cir. x R.P.M / S.F. per Min.) 12 | Bright White Fan-Shaped Burst | 1. Nickel-Black Shape close to wheel. 2. Moly-Short Arrow Shape Tongue (only). 3. Vanadium-Long Spearpoint Tongue (only). | Red Carrier Lines (Very little carbon exists) | Long Straw Color Lines (Practically free of bursts or sprigs) |

**Figure 11-1** Identification of metals. (*Courtesy of Hobart Brothers.*)

The majority of the steel being welded in the repair situation does not have the steel classification marked or stamped on the metal to be repaired. Other methods must be used to identify as accurately as possible the type of steel or iron. Please refer to the identification of metals chart (see Figure 11-1); different types of steels and iron are listed and the various methods of establishing the identification of the metal are shown. The spark test is one of the most commonly used methods. The spark test method is the observing of sparks caused by friction by holding a bar of metal lightly against an emery wheel.

NOTE:   Use the spark test in darkened areas or against a dark background for better visibility.

Another method used to check the degree of hardness of the steel is by use of the file test. By passing a common mill file over a piece of steel, the degree of hardness can be determined. For example, on mild steel, the file will remove the metal easily. On steel with medium carbon content, the file will cut using more pressure. Tool steel requires greater pressure, whereas hardened tool steel cannot be cut and will dull the teeth of the file.

Using the right electrode as filler metal for the type of steel involved can mean a satisfactory weld. The wrong filler metal or electrode can lead to failure of the weld metal or the work piece being welded.

## QUESTIONS

1. What organizations have adopted a uniform steel classification?

2. What does the first letter represent?

3. What does the classification number 1030 mean?

4. How can steel or iron be identified?

5. If one piece of metal files easily and the other piece will dull the file, what types are they?

# TYPES OF WELDS AND JOINTS

## chapter 12

Every welding job or operation will contain one or more of the following welds and beads: surface weld, groove fillet, plug, rivet, and slot.

## WELDS

A *bead* weld is a narrow layer of weld metal deposited in a straight unbroken line either on a flat surface (see Figure 12-1a) or deposited on a closed (Figure 12-1b) butt weld. A bead weld can also consist of two or more layers of

deposit weld metal called padding. A bead weld is often called a stringer bead.

A *groove* weld is a weld where weld metal is deposited in a groove, which can consist of one or more beads depending on the type of groove and the thickness of the metal. The groove is formed by an open square butt joint (see Figure 12-2a), a bevelled butt joint (Figure 12-2b) or V (see Figure 12-2c), a U-joint, or an outside corner joint (Figure 12-2d).

A *fillet* weld is a weld where weld metal is deposited in one or more beads in the angle

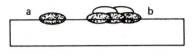

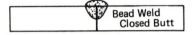

**Figure 12-1** Bead weld: (a) single; (b) side by side or multi-layer.

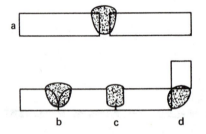

**Figure 12-2** (a) Groove weld open butt; (b) bevel or V joint; (c) U joint; (d) outside corner.

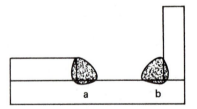

**Figure 12-3** Fillet weld: (a) lap weld; (b) inside corner.

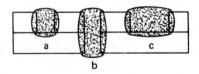

**Figure 12-4** (a) Plug weld; (b) rivet weld; (c) slot weld.

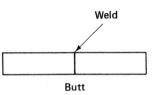

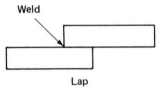

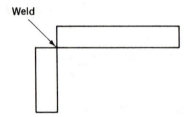

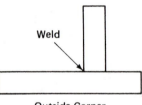

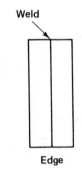

**Figure 12-5** Types of joints.

formed by two plates such as a lap weld, (see Figure 12-3a), inside corner weld (Figure 12-3b), or an open outside corner and bevelled outside corner joint.

The *plug, rivet,* and *slot* welds are very similar because they join two overlapping pieces of metal. The plug weld (see Figure 12-4a) is where a hole is drilled into the upper plate and the two pieces are welded together by welding in the hole. The plug is sometimes called a button hole used for welding light gauge metal. The rivet weld (see Figure 12-4b) is where the hole is drilled through both upper and lower pieces of metal and the weld is made clear through the hole. A variation of this type is when the hole in the upper piece is larger than the lower piece. A slot weld (see Figure 12-4c) is where the weld metal is deposited into an oblong slot instead of a round hole. The slot is generally used for fastening down deck plates.

The plug, rivet, and slot welds are not used as often as the bead, groove, or fillet weld; and they are comparatively inefficient.

## JOINTS

There are five basic types of joints used in arc welding. They are butt, lapp, inside corner or T-joint, outside corner, and edge joint (see Figure 12-5). The welder must decide which is the best type joint to use, which is the strongest and the weakest under certain stress or load conditions. Some other factors to consider are: can the type of joint selected be welded easily, what is the cost of preparing the joint, and how much material will be used.

A certain type structure can be made with less material, less time, and be stronger, with the proper design rather than one that was improperly designed.

## QUESTIONS

1. What is a bead weld?
2. What is a groove weld?
3. What is a fillet weld?
4. What is a plug weld?
5. What are some of the most common joints used?

# JOINT

# PREPARATION

## chapter 13

After a suitable joint has been selected for a particular job, the next step or thing to consider is how can the proper penetration and fusion be obtained. The penetration should be as near 100% as possible or else the joint will be weaker than the base metal and possibly fail.

The thickness of the metal, in many cases, determines what joint preparation is necessary. For example, when butt welding two plates together, if the metal is less than 3/16″ (4.76 mm) (see Figure 13-1), there will be adequate penetration in one pass or bead weld when the plates are spaced apart about 1/16″ (1.58 mm) or about half the thickness of the plate, allowing for expansion.

For plates greater than 3/16″ to above 1/2″ (9.52 mm to 12.7 mm) in thickness (see Figure 13-2), a single groove or a V-groove will be necessary, depending on the thickness, to get the proper penetration. For plates over 1/2″ (12.7 mm) in thickness, a double V- or J-joint (see Figure 13-3) is preferred to a single V to reduce the possibility of distortion because it can be welded on both sides of the plates. The

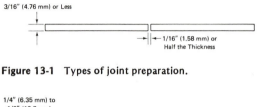

Figure 13-1 Types of joint preparation.

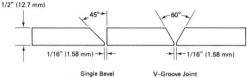

Figure 13-2 Types of joint preparation.

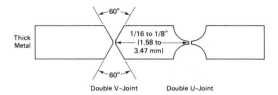

Figure 13-3 Types of joint preparation.

plates can be prepared by several methods depending on the thickness and the type of equipment available.

The neatest type joint preparation is by using a grinder with a cup type stone or a depressed center grinding wheel. On heavy plate, this method may be too time consuming and may be expensive in respect to material used.

Another very common method is the use of the oxy-acetylene cutting torch. The procedure for cutting is described in Chapter 4 under cutting bevels. After the metal has been cut, the cut area should have all of the slag removed, especially by hammering or chipping

in the bottom edges and may be touched up a little with the grinder to avoid getting slag particles in the weld metal and causing a weak spot. There are special electrodes available for preparing joints or regular electrodes (see Chapter 29) for cutting metal may be used, but the slag must be removed for a clean joint.

The air carbon arc can be used for joint preparation. The metal is melted by the heat of the arc between a carbon electrode and the base metal. As the metal melts, a jet of compressed air blows out the molten metal. The air jet is built into the electrode holder and directs the air to the molten metal. The air is controlled by a valve built in the handle.

Whatever method of joint preparation is used, before welding any joint, most of the dirt, paint, slag, or any foreign material should be removed to prevent impurities from getting into the weld metal.

## QUESTIONS

1. What determines the type of joint preparation?

2. If two plates under $\frac{3}{10}''$ (7.75 mm) are spaced apart half the thickness and are to be butt welded, what type of joint preparation is necessary?

3. What type joint preparation is necessary for plates under $\frac{1}{2}''$ (12.7 mm) thick?

4. What about plates over $\frac{1}{2}''$ thick?

5. What are some methods used for bevelling plates?

# WELDING SYMBOLS

## chapter 14

In the previous discussion on the electrode numbering system, we noted that each letter, number, or symbol represents a certain characteristic of the electrode—the number system is a form of shorthand in describing the electrode, standardized by the AWS and ASTM.

The same has happened to welding symbols that are used by draftsmen, designers, welders, and others involved in the making or reading of blueprints or drawings.

The symbols describe how the structure or object is to be made, the type of welds necessary, where to weld, joint preparation, and other important information. Before a standardized system was used, it was often the welder's decision or responsibility to make the proper welds. Now, the necessary information is furnished on the print or drawing for the welder to follow (see Figure 14-1). However, there are many welders in the field or in the repair business who never use or are not required to use or read prints with welding symbols.

The welding symbol consists of a base or reference line with an arrow on one end pointing to the weld area (see Figure 14-2). At the junction of the reference line and the arrow there may be a flag or solid dot indicating a field weld. This means that the structure is to be finished after the removal from the shop.

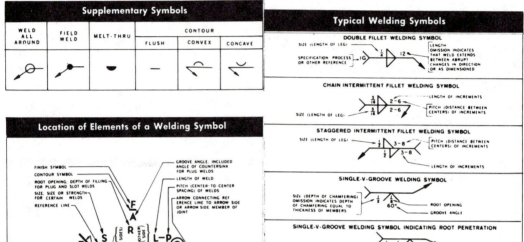

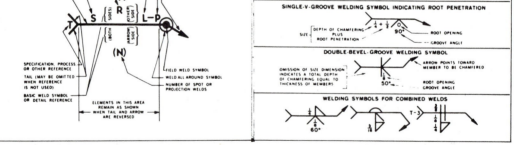

**Figure 14-1** AWS welding symbols. (*Courtesy of Lincoln Electric.*)

No mark means that it is welded in the shop. A circle at the same point is the symbol to weld all around. The symbols that are above or below the base line indicate the type of weld to be used, on what side of the structure, contour and size of bead, length of weld, and the type of welding process. When a geometric figure such as a triangle is used for fillet welds, this is on the top of the reference line, and the weld is made on the opposite or other side from the

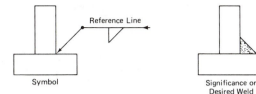

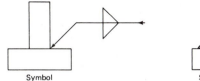

**Figure 14-2** Symbol. Significance or desired weld.

**Figure 14-4** Symbol. Significance or desired weld.

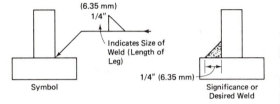

**Figure 14-3** Symbol. Significance or desired weld.

Other figures represent different types of welds such as spot, seam weld and types of joints such as bevels, V-joints, and others.

When fractions appear with the weld symbol, that indicates the size of the bead. Numbers on the line specify the length of the beads or welds and the distance betwen centers of the welds. At the end opposite the arrow there is a V-shaped line used for indicating specification type of electrodes, processes, or other important information.

arrow (see Figure 14-3). If the figure is below the line indicated, the weld is made on the same side as the arrow (see Figure 14-2). When the figures are above and below the line, it means that the piece is welded on both sides (see Figure 14-4).

Symbols are complicated in the beginning, but after the welder uses them for a short time, they become quite simple.

## QUESTIONS

1. What is considered to be welding shorthand?

2. Are the welding symbols standardized?

3. Why are symbols used?

4. What is the line called where the symbols are placed?

5. If a dot or flag appears at the junction of the arrow and base line, what does it mean?

6. If the triangular figure is on top of the base line, where is the weld to be made?

7. When a fraction appears in the symbol, what does it mean?

8. What information appears at the opposite end of the base line?

# ARC-WELDING: CORRECT WELDING PROCEDURES

## chapter 15

There are several important points to remember in order to obtain a good weld besides using the proper electrodes. Use the proper size electrode (diameter) in relation to the type and position of weld, the thickness or mass of the metal, joint preparation, the ability of the electrode to carry high current, and its effect on the original properties after welding.

Normally, an electrode should never exceed the thickness of the work or base metal, such as welding steel plate. However, a situation may arise when smaller electrodes are not available at the moment. Care must be used not to burn holes in the metal. This can be accomplished by adjusting the amperage and travel speed.

Using larger electrodes permits faster travel thus saving time.

Current is an important factor because if the current is set too high, the electrode will burn off too fast with a large and uneven bead; this may result in excessive penetration. If the current is too low, there will not be enough penetration because there is not enough heat to melt the base metal. In addition, the bead will be too small, too high, and very uneven. Study Figure 15-1 for good and bad beads.

The correct arc length is imporant because if it is too short or close, the voltage will be too low with not enough heat to melt the base metal for adequate penetration. The bead will

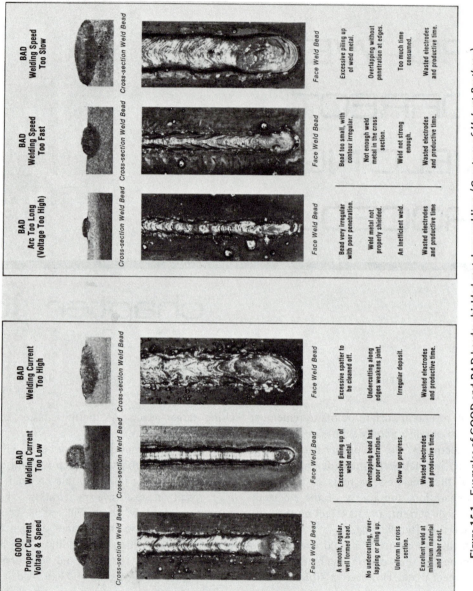

**Figure 15-1** Examples of GOOD and BAD beads: shielded metal arc welding. *(Courtesy of Hobart Brothers.)*

be too high and uneven, and the electrode may stick to the metal. If the arc length is too long, the voltage will be excessive, causing the arc stream to wobble. The metal from the electrode will melt in globules causing a wide, irregular bead and excessive spatter. There will be decreased fusion between the base metal and the weld metal.

One of the biggest problems with inexperienced welders is the speed of travel. If the speed is too fast, the bead is too narrow, leaving pointed ripples with the pool of molten lead solidifying too fast with impurities being trapped inside. These beads are too high and lack the proper penetration. If the speed is too slow, the beads are too wide, with excessive metal deposited and excessive penetration with holes burned in the metal if thin plates are being welded.

The correct electrode angle is important when making fillet welds and groove welding. When using a heavy flux-coated electrode in fillet and lap welding, the molten slag could present a problem by getting in the middle of the bead. This prevents the molten metal from flowing together between the two plates if the electrode movement or manipulation is not correct (see Figure 15-2). Also, if the angle is too high, undercutting of the vertical plate could result. Undercutting of the vertical plate could cause a weak spot, depending on where the plate is used—for example, if the plate is used for a supporting brace (see Figure 15-3). However, in the repair field, many times the correct electrode angle cannot be maintained because of the location of the weld area or the type of work or structure to be welded.

NOTE: When arc welding, always watch the formation of the bead at the edge of the molten pool of metal. This is a guide for a good bead with the proper contour and ripples. The ripples should be smooth and rounded.

# PROCEDURE

## *SAFETY CHECK BEFORE WELDING*

Follow safety precautions discussed previously, such as checking the area for fire hazards, locating fire extinguishers, and making sure the area is free of litter and scrap. Check welding equipment such as cables, electrode holder, good ground clamp, helmet and lenses, clear work table, and cable connections. If there are any defects, correct them before starting to weld. Personally check for proper clothing, pants, cuffs, buttoned shirt pockets, proper shoes, clear pair of safety glasses; put on protective clothing, if available; and remove matches or inflammables from your pockets.

## *WELDING*

Select several pieces of plate steel 3/16" (4.76 mm) or thicker, 6" X 6" (152 mm X 152 mm) or larger, 1/8" (3.17 mm) E6012 or E6013 electrodes, and check the amperage

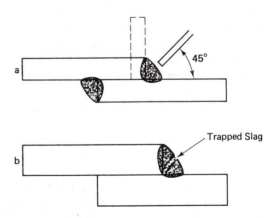

**Figure 15-2** (a) Proper angle for lap and fillet welding; (b) slag inclusion due to improper electrode movement or manipulation.

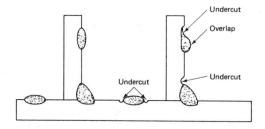

**Figure 15-3** Correct beads and undercut beads.

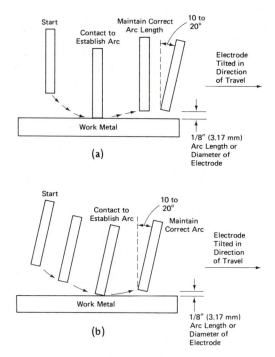

**Figure 15-4** (a) Tapping method to start arc; (b) scratching method to start arc.

setting. Many machines work well from 90 to 100 amperes. (The E6012 or E6013 electrodes will have a more defined bead than E6011 or 10, and will be easier to check progress.) Make sure you have a chipping hammer and wire brush.

Position a plate on the bench, remove dirt and rust with wire brush. Place an electrode in the holder with the bare end in the jaw of the holder.

*CAUTION: Do not turn the machine on.*

Practice using one of two methods to strike an arc; tapping or scratching (see Figure 15-4). The scratch method is scratching the electrode along the surface of the metal, the same as striking a match. Tap method is striking the plate with an electrode at a 90° angle then moving the electrode upward about 1/8″ (3.17 mm) from the surface. Try one or both of these methods to get used to the feel of the holder. Do not grip the electrode holder too tightly because it becomes tiring after a short time. Connect the ground cable to the bench and check for good contact. Put on the helmet and check for headband adjustment. It should be tight enough to stay on your head when the helmet is raised, but it should not be uncomfortable. Put on gloves, if available. Pick up the electrode holder and electrode and turn on the machine. Be sure you know the location of the switch in case of an emergency.

Lower the electrode to 1″ to 2″ (25.4 mm to 50.8 mm) from the surface of plate and then

lower your helmet. Strike the electrode on the plate with either the scratching motion or tapping motion to establish an arc. Don't move too fast or else the arc will stop. As the arc starts, raise the electrode up a little to about 3/16″ (4.76 mm) (long arc) for a moment to establish the arc and form a molten pool of metal. Then lower electrode to about 1/8″ (3.17 mm) from the surface. After a few seconds, move the electrode up to break the arc. Repeat striking or starting the arc using both methods at least ten times.

NOTE: If the electrode sticks to the surface, twist the electrode and break it loose. If still stuck, release electrode from holder or shut the machine off, but keep the helmet down over your face.

After you practice striking an arc, the next step is running short beads.

Strike the arc and maintain the arc moving forward. If you are right-handed, move the electrode from left to right. If you are left-handed, move the electrode in the opposite direction. This way, you can observe the formation of the puddle and bead. Keep the electrode at right angles or vertical to the plate and tip the electrode at about a 10° to 20° angle forward at the tip. Move in a straight line for a distance and stop the arc. Chip off the slag with the chipping hammer and observe the results

*CAUTION: Be sure to wear the clear safety glass or chipping goggles.*

Check results with Figure 15-1, good and bad beads.

## QUESTIONS

1. How important is the current setting?
2. How important is the arc length?
3. What causes narrow pointed beads?
4. What causes excessively wide beads and penetration?
5. What causes undercutting on vertical plates?
6. Why is it important to check the area before starting to weld?
7. Why should E6013 electrodes be used for practice instead of E6011?
8. What are the methods of starting the arc?
9. What should be done if the electrode sticks and cannot be released?
10. What angle should the electrode be in the direction of travel?

# RUNNING BEADS

## chapter 16

### SHORT BEADS

Turn the machine off. With a piece of soapstone, draw a series of lines 3″ to 4″ (76.2 mm to 101.6 mm) long and about 1″ (25.4 mm) apart, on a 4″ × 6″ (101 mm × 152 mm) plate (see Figure 16-1). Then weld a bead on each line, using 1/8″ (3.17 mm) E6013 electrodes; keep the bead straight and on the line. Keep observing the bead formation at the back of the molten puddle. Chip off the slag and observe the shape of the bead, the height, width, and the type of ripples produced (check with Figure 15-1). Is the width and height of the bead uneven? Burn off each electrode to a stub of less than 2″ (50.8 mm) remaining (any-thing more is wasteful) and put the stubs into a can. Do not throw them on the floor—this will prevent your slipping on them. Try several pieces, but make the bead longer in length, spacing 1/2″ (12.7 mm) apart. The only way to gain experience and develop the proper technique is practice. Practice on beads before attempting to make joints. Master one step at a time.

### UNDERCUTTING AND OVERLAPPING

Undercutting can weaken a welded area especially when welding joints, but it will occur when welding flat beads (see Figure 16-2a, also

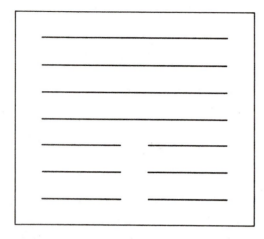

Figure 16-1 Practice beads: Short and continuous beads patterns. Use soapstone or chalk. Use both sides of plate.

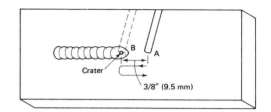

Figure 16-3 Restarting bead and filling crater. Start arc about 3/8" (9.5 mm) ahead of crater at A, move back to crater B. Using a longer arc, fill crater to proper height, then move forward, lowering electrode to proper arc length.

Figure 15-3). In flat beads, undercutting will leave a groove along both sides of the weld; and it is caused by using an excessive amount of current. Undercutting is also caused by the incorrect angle of the electrode on the vertical plate, which will be discussed in fillet welds.

Overlapping is metal deposited in a weld that has not fused to the base metal along the outer edges of the weld (see Figure 16-2b). This condition is caused by the current being set too low.

### RESTARTING THE BEAD

The length of the bead may be interrupted by sticking the electrode or changing electrodes. To restart the bead, chip the slag from the end of the bead and brush away. Restart slightly ahead of the existing bead (see Figure 16-3) and

return to the crater using a longer arc. Fill in the crater with sufficient metal, then lower the electrode to the proper distance and continue the bead.

Craters generally form at the end of the completed bead. This could cause failure if under load pressure. To eliminate the crater, pause a moment at the end of the bead to fill the crater, using a short arc, then quickly back or whip the electrode backwards to break the arc (see Figure 16-4).

## KINDS OF BEADS

### WELDING CONTINUOUS BEADS

Using the same procedure as before, draw a series of lines, using soapstone, 1/2" (12.8 mm) apart on another plate (see Figure 16-1). This

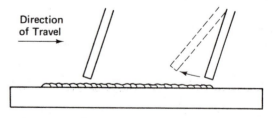

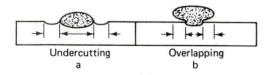

Undercutting
a

Overlapping
b

Figure 16-2 Incorrect or defective beads.

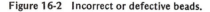

Figure 16-4 To eliminate craters at the end of a bead, shorten arc length, pause to fill in end of bead, then move electrode back over the bead, lengthening the arc gap to break it or whipping the electrode to break the arc.

time, use either 1/8" (3.17 mm) E6010 or E6011 electrodes and reset current according to the manufacturer's recommendations, or between 75 and 120 amperes. Some electrodes work well at 75 and others at 90 or 100 amps, depending on the machine and make of electrode. After running the beads, remove the slag and check the appearance of the beads as compared to those of the E6013 electrode. Check for contour, width, and the type of ripples.

NOTE: Another electrode, E7014, is excellent for practice because of its well-defined bead and easy slag removal.

## MULTI-DIRECTIONAL BEADS

With a plate larger than 6" X 6" (152 mm X 152 mm), draw a line with soapstone 3/4" (19 mm) from the edge, around the plate, within 3/4" (19 mm) of the end of the plate where the line started (see Figure 16-5). Continue the line within 3/4" (19 mm) inside the line toward the center. Continue until the line is in the center of the plate. Use E6013 or E7014 electrodes and follow the recommended current setting. Weld the full length of the line. This is to develop skill in welding in any direction. Remove slag and observe the bead. Using the same type procedures, change the electrode to the

E6011 electrode and follow the recommended correct setting. Remove the slag and observe.

## WEAVING BEADS

Sometimes, it is necessary to produce a wider bead—especially when welding a poorly fitted joint, where buildup is required, or when welding in different positions. In many cases, it is better to make one wide bead rather than two or three narrow beads. The wider bead can be made by using the *weaving* technique, which is the sideways motion of the electrode while moving forward. Some of the more frequently used weave beads (see Figure 16-6) are: Box, Crescent or open U, Figure 8, and Zig Zag or simple side to side.

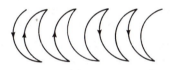

Cresent of Open U Weave

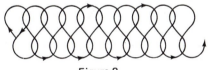

Figure 8

Box Weave

Zig Zag or Side to Side

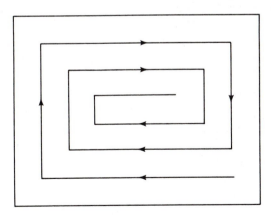

**Figure 16-5**  Multi-directional bead pattern.

**Figure 16-6**  Types of weaving.

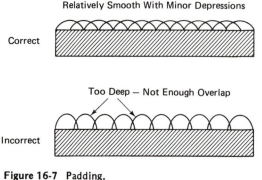

Figure 16-7 Padding.

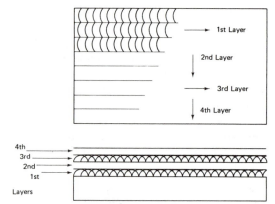

Figure 16-8 Padding pattern.

In the following sections, additional information will be given as to the correct type of weave bead to use for the specific joints.

Using the same procedures for bead, draw lines 3/4" (19 mm) apart on a plate about 6" X 6" (152 mm X 152 mm). Weld beads on the lines using one of the above types of weave motions. The width of the bead should be about three times the width of the electrode. In some cases, the beads would be wider, such as in padding. Remove the slag and check the contour and width, which should be even on the sides of the bead. Use the other types of weaving motion to further develop the welding technique.

## PADDING BEADS

Many times, parts of machinery are worn and it is cheaper to repair than replace, or the parts are not readily available. Heavy machine shafts can be padded and machined on a lathe, adding life to them. Padding is the process of building up a surface by depositing a series of overlapping beads (see Figure 16-7) which can be regular or weave type beads.

Depending on the situation, a larger electrode such as 5/32 electrode rather than a 1/8" (3.17 mm) electrode should be used. Each bead should overlap the preceding one by 1/5" to 1/4" to avoid valleys or depressions be-

tween the beads. When the first series of beads is finished, the surface should be fairly smooth.

Use the 1/8" (3.17 mm) E6013 electrode, and follow the manufacturer's recommended current setting. Select a plate 1/4" X 4" X 6" (6.35 mm X 101.6 mm X 152 mm). Weld a bead along the long edge of the plate. Chip off the slag and thoroughly wire brush to remove all traces of slag. Weld a parallel bead alongside of the first one overlapping it by 1/4" to 1/5". Remove the slag and wire brush. Check the depression between the two beads. If the depression or valley is too deep, try overlapping the adjacent bead by 1/3". Continue to weld beads the full width of the plate. Try alternating regular beads and weave beads. Check for smoothness and depressions.

NOTE: Remember to remove the slag after each bead to avoid having slag trapped in the metal.

Weld the next layer of beads at right angles with the first layer with the same procedure as the preceding layer, with regular and weave beads. Weld the third layer at right angles with the second layer (see Figure 16-8), with the same procedure as the preceding layer. Add the fourth layer. Notice the improvement in welding technique with each additional layer.

# QUESTIONS

1. Where should you observe the bead formation?

2. How long should the electrode be before you discard it?

3. What is undercutting in flat beads?

4. What causes undercutting?

5. What is overlapping?

6. When restarting a weld or bead, where should the arc be struck?

7. Where are craters generally found?

8. What is another electrode that leaves a well-defined bead and slag that is easily removed?

9. What is the purpose of the multi-directional beads?

10. What is the purpose of the weave bead?

11. What is padding?

12. How should the 2nd layer be applied?

# BUTT JOINT WELDS

## chapter 17

The butt joint is where the edges of two pieces of metal are in line or in the same parallel plane together. When welded, the two pieces form a flat surface. Some of the most common types of butt joints are found in storage tanks, boilers, and refuse bodies. This type of joint must have 100% penetration of the joint for maximum strength. For plates that are less than 1/8" (3.17 mm) thick, sufficient penetration can be accomplished by spacing the plates apart equal to one-half the thickness of the plates (see Figure 17-1). This spacing is also recommended to prevent distortion from expansion and contraction of the welded joint (Figure 17-2). Welding plate over 1/4" (6.35 mm) thick must have the edges beveled to a 30° angle

(Figure 17-3) and must leave a 1/16" (1.58 mm) to 1/8" (3.1 mm) root face or land on the bottom of the plate (see Figure 17-4).

Root of a weld is the part of the weld which is farthest from the application of the weld heat.

If the opposite side of the plate is accessible (and for added strength), a double bevel or double V-bevel can be used. When bevelling a joint, an excessive angle such as a 45° can cause a problem because as the weld contracts, it will be excessive, causing warping. On thicker metal, more than one pass or bead must be made. The first one at the root area would be called a root or penetration weld (see Figures 17-4 and 17-5).

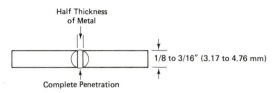

Figure 17-1  Open joint.

The second one, if needed, could be called the filler; the top one is generally called a finish or cover weld. The second and third would be a weave bead on heavier metal.

Distortion can be a major problem when welding thin plates with a long continuous bead. This is called lengthwise distortion; the plates bend upward in the direction of the weld (see Figure 17-6). Clamping the pieces to a workbench would help prevent distortion but many times this is not possible. The use of skip weld or backstep weld techniques will help to eliminate distortion.

Skip welding is welding a series of beads, starting about 2" (50.8 mm) from the end of the plate and welding back to the end (Figure 17-7). Next bead, start about 6" (152.4 mm) from the end and weld backwards for 2" (50.8 mm). Continue this procedure to the end of the plate. Then, after cleaning weld, start to fill in between the beads.

To backstep, start about 2" in from the edge and weld back to the end of the plate. Clean weld and start about 4" in and weld backwards to the first bead. (Figure 17-8). Continue this procedure to end of the plate. Excessive amperage will cause the molten pool to become too large, which will cause excessive contrac-

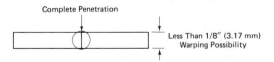

Complete Penetration

Figure 17-2  Closed joint.

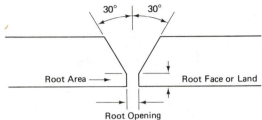

Figure 17-3  Beveled plate.

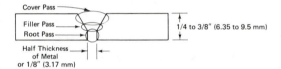

Figure 17-4  Single V or groove.

tion when cooling. If the plates are too tight together when welded with high amperage and large molten pools of metal, the weld could crack due to expansion and contraction (for another problem, see Figure 17-9).

## BUTT WELDING—THIN PLATE

Use E6013 or E6012 1/8" (3.17 mm) electrodes and set welder to recommended current. Place two plates 1/8" X 2" X 8" (3.17 X 50.8 X 203 mm) about 1/16" (1.58 mm) apart spacing about 1/2 the thickness of the plate but not exceeding 2/3 of the thickness of the plate (see Figures 17-10 and 17-11). Tack weld the plates about 1/2" (12.7 mm) from each end. Weld a continuous bead across the length of the plate.

Figure 17-5  Double V or groove.

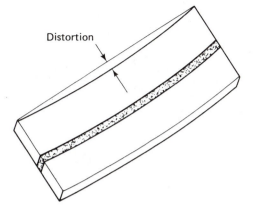

Figure 17-6 Lengthwise distortion caused by a continuous weld.

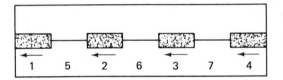

Figure 17-7 Skip weld method. Controlling lengthwise distortion.

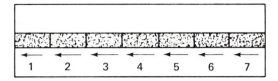

Figure 17-8 Backstep method. Controlling distortion.

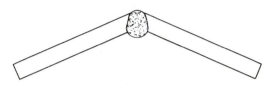

Figure 17-9 Distortion caused by contraction when metal cools and excessive molten metal is in weld pool.

Make sure to keep the electrode in the center of the space between the two plates. Remove slag after welds have cooled. Cut the welded plate in half, across the weld. Check for penetration of the weld between the plates. Place one of the pieces in a vise and using a large hammer, bend it back and forth until it breaks. Check for slag inclusion porosity, penetration, and incomplete fusion (another method, use a tapered gap, see Figure 7-12).

Using an E6011 1/8" (3.17 mm) electrode, reset welder to recommended current, and repeat procedure using two plates. After the plate has cooled, remove the slag. Cut the welded plate in half across the weld. Check for penetration of the weld between the plates. Place one piece in a vise with the weld slightly above and parallel with the jaws of the vise (Figure 17-12a). Break as before and check for slag inclusion, porosity, penetration, and incomplete fusion. Repeat this procedure with increased current setting and decreased current.

## SINGLE GROOVE, V, OR BEVELLED BUTT WELDS

Use 1/8" (3.17 mm) E6011 and E6013 electrodes. Select two plates 1/4" to 3/8" (6.35 mm to 9.5 mm) thick × 2" × 8" (50.8 mm × 203 mm). Bevel one long edge of each plate to 30° leaving a 1/16" (1.58 mm) to 1/8" (3.17 mm) root face or land (see Figure 17-4), using a grinder or cutting torch. Position the plates with a 1/16" (1.58 mm) to 1/8" (3.17 mm) space between the two plates as indicated in illustration. Using a E6011 electrode,

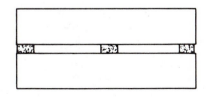

Figure 17-10 Tack weld to control distortion.

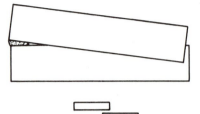

**Figure 17-11** Contraction causes metal to overlap.

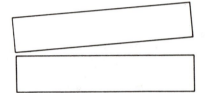

**Figure 17-12** A tapered gap can be used.

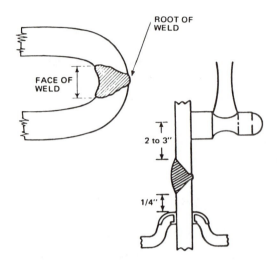

**Figure 17-13** Method of testing welds. (*Courtesy of Airco*.)

set the welder to the proper current and tack weld the two pieces together about 1″ (25.4 mm) from each end. Run a bead along the bottom of the groove and be sure that both edges are evenly penetrated. This is called a root or penetration weld (Figure 17-4). Remove slag and clean thoroughly. Deposit a second bead using a slight weaving motion covering the width of the groove. If a third (or more) bead is required, depending on the thickness of the metal, clean the weld before each new weld. The cover or finish weld requires a wider weaving motion with the edge of the weld being slightly wider than the width of the groove. The weld should be slightly higher than the surrounding area. Remove slag and clean weld. Observe the contour, evenness of the weld edges, and the root weld penetration. Cut several pieces off the plate about 2″ (50.8 mm) wide. Observe the penetration of the welds and possible flaws. Bend one piece in press and check for lack of fusion at root weld and slag

inclusion. Repeat the procedure using E6013 electrodes. Repeat with different thicknesses of metal.

## DOUBLE BEVEL OR GROOVE WELDS

When welding plates 1/2″ (12.7 mm) thick or heavier, it is desirable to use a double bevel method if the both sides of the plate are accessible (see Figure 17-5). The root or land then would be in the center of the plate for proper penetration. Alternate welds, put one weld on one side and then one weld on the opposite side. This will eliminate the possibility of distortion that could arise if one side of the groove or bevel was welded first, then the other side. Also, after the root weld is completed, a larger electrode could be used, such as an 5/32″ (3.96 mm) electrode to save time. With heavier plates, more welds or passes must be made to fill the groove with proper cleaning before each weld to avoid weakening the joint.

# QUESTIONS

1. What is a butt joint?
2. How much penetration is necessary for maximum strength?
3. What is meant by the root of the weld?
4. What is lengthwise distortion?
5. What type welds can be used to help eliminate distortion?
6. What is the approximate spacing of the plates?
7. What is the angle of the bevel on each plate?
8. On heavier plate, is it desirable to bevel both sides?

# LAP JOINTS

## chapter 18

The lap joint is one of the most frequently used joints in welding. It is also economical because it requires very little joint penetration since bevelling is not required. The amount of overlap depends on the strength requirements, the thickness of the metal, and the direction of stress on the joint. For the greatest strength, both edges of the joint should be welded, if possible. But in many cases, only one side is accessible, such as joining sections of some car frames. The lap joint is used to reinforce various types of structures and in the repairing of heavy equipment.

## PROCEDURE

Select two steel plates, 1/4" X 2" X 8" to 12" (6.35 mm X 50.8 mm X 203 mm to 304 mm). Use 1/8" (3.17 mm) E6011 and E6013 electrodes. Set up the plates with the upper plate overlapping the lower by about 1" (25 mm). Check the contact between the two plates; if a large gap is evident, straighten them. First, use the E6011 electrode and set welder to recommended current. Start the bead with the electrode at a 45° angle with the plate and tip

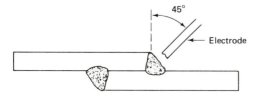

**Figure 18-1** Lap joint single bead.

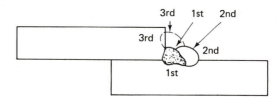

**Figure 18-2** Lap joint multi-bead or weld.

the electrode holder about 20° to 30° in the direction of travel (see Figure 18-1). Use a slight weaving motion to create a bead about 1/4" (6.35 mm) wide. Make sure there is complete fusion at the root and avoid overlapping the top edge. After some cooling, weld the other lap joint on the opposite side. Remove the slag and observe the bead for contour and fusion of the lower and top plate. Check for undercutting of the lower plate. Repeat the procedure using the E6013 electrode and observe. Also, weld the other lap joint. If the top edge is starting to melt away, direct the arc toward the bottom plate, especially if the top plate is thinner.

## LAP JOINTS, MULTIPLE PASS

When welding plates 3/8" (9.52 mm) thick or thicker, it is impossible to weld a strong joint with one pass or bead. Two or more passes or beads are recommended (see Figure 18-2). When using two passes, the first bead should be made with very little weaving with complete fusion of the root. Thoroughly clean the bead to avoid trapping the slag between the two beads and causing weakness of the joint. The second pass or bead should be made with a weaving motion with a slight pause at the top of the weave to fill in the vertical portion of the joint.

**Figure 18-3** Slag inclusion.

On joints involving heavier metal, three or more passes or beads are necessary. The first one is the root bead with complete fusion of the root. The second pass or bead covers most of the root bead and completes fusion with the lower plate. The third pass or bead covers the remainder of the root pass and half of the second pass with complete fusion with the top plate. The second bead serves as a shelf for the third bead. If more passes are required to properly fill the joint, follow the same procedure as the second and third bead. The finished bead should have a 45° face or surface. The angle of the rod is altered to the various conditions encountered. In many cases of repair jobs, the recommended angle cannot be maintained because of the accessibility of the area to be repaired.

NOTE: Slag inclusion may be a problem for inexperienced welders (see Figure 18-3). This is often found when using E6013 and E7014 electrodes. Slag becomes trapped in the root of the weld thus preventing the fusion of the top and lower plates. The two common causes are: holding the arc too short or too close and improper electrode manipulation.

## QUESTIONS

1. How much joint preparation is required for lap joints?
2. How can the greatest strength be obtained in lap joints?
3. How can melting away of the top edge be avoided?
4. Why is it necessary to clean the first bead before welding the second bead?
5. What is slag inclusion?

# FILLET WELDS
# (Inside Corner or
# T-Welds)

# chapter 19

The fillet weld is similar to the lap weld. It is very commonly used in fabrication of various structures. If stresses are placed on the opposite direction of the welded joint, the joint is too weak. By adding a weld on the other side of the joint, this adds sufficient strength to the joint.

A common defect in fillet welding is the undercutting of the vertical plate, which could cause failure under stress. This is caused by several reasons but the most common ones are the incorrect electrode angle and improper electrode manipulation.

## SINGLE PASS

Use four plates 3/16″ or 1/4″ X 3″ X 8″ to 12″ (4.76 mm or 6.35 mm X 76 mm X 203 mm to 304 mm), 1/8″ (3.17 mm) E6011 and E6013 electrodes. Set the vertical plate in the center of the horizontal plate and tack weld both ends of the vertical plate to the horizontal plate. Use the E6011 electrode and set the welder to the recommended current. Run the weld similar to that of the lap weld with the electrode held at a 45° angle with the electrode holder tipped forward in the line of travel (see Figure 19-1). Observe the bead formation and the proper fusion of both plates at the root of the weld. Little or no weave motion with the proper speed of travel will deposit enough metal to form a 1/4″ (6.35 mm) bead. If undercutting of the vertical is evident, concentrate the arc more toward the horizontal plate. After sufficient cooling, remove the slag and observe the bead. On the opposite side, use the E6013 electrode with the recommended amperage and run a bead. This electrode will cause more problems with slag inclusion than the E6011 electrode. Repeat the above procedure, using more of the E6013 electrode than the E6011. Save the welded plates for the following exercises.

## FILLET WELDS USING MULTI-PASS

To obtain a stronger inside side corner or T-weld, especially when using heavier plate metal, a series of beads or layers of welds can be used. This may be the case where a strong weld is required and only one side is accessible.

Use a 1/8″ (3.17 mm) E6013 electrode, 5/32″ (3.96 mm) E7024 electrode, a welded T-joint plate. Carefully remove the slag from one of the welds. Using an E7024 electrode, set the welder to the recommended current. Run a bead on top of the existing bead using a slight weaving motion (see Figures 19-2 and 19-3). Be sure to obtain adequate fusion with both plates and even distribution of the weld metal on both plates. The face of the bead or weld should be about 3/8″ (9.52 mm). Remove the slag and run a bead on top of the second bead with the proper fusion of both plates with a wide weaving motion. Remove the slag and check the contour of the bead. The weld metal should be evenly distributed on the vertical and hori-

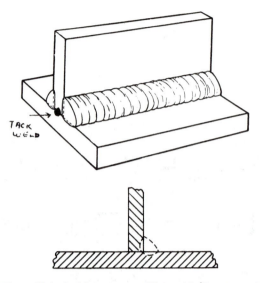

**Figure 19-1** Inside corner or fillet weld. (*Courtesy of Airco.*)

zontal plate. On the reverse side of the plate, remove the slag. Using an E6013 electrode and welder set to proper amperage, run a bead covering the majority of the existing bead or weld similar to that of the lap weld. Hold the electrode at about a 70° angle from the horizontal plate and use a slight weaving motion. But stop the bead within about 1½″ (38 mm) from the end. Remove the slag and run a third bead, completely covering the first bead and about half of the second bead using an electrode angle of about 20° from horizontal, with proper fusion with the vertical plate. A slight weaving motion is better to fuse the vertical to the beads. End bead about 2½″ (63.5 mm) from the end. Remove slag and observe the distribution of the weld metal and the evenness of the beads. Check for undercutting of vertical plate. Any number of beads can be made with each series starting at the horizontal plate with proper fusion with the existing beads.

*CAUTION: Be sure to remove the slag after each bead.*

ARC SHORT — SPEED CONSISTENT — NO OSCILLATION

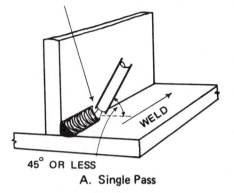

45° OR LESS

A. Single Pass

**Figure 19-3** Example of multi-pass fillet weld: 1st bead straight and 2nd bead weaving. (*Photo by S. Suydam.*)

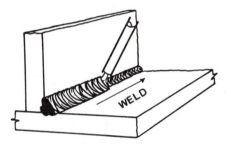

B. Second Pass

**Figure 19-4** Vertical plate is out of alignment or distorted due to contraction of the weld metal.

WELD

INVERTED
U — WEAVE

BOX WEAVE

CIRCULAR
WEAVE

C. Weave Patterns

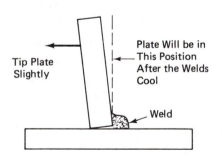

Tip Plate
Slightly

Plate Will be in
This Position
After the Welds
Cool

Weld

**Figure 19-2** Multi-bead fillet weld. (*Courtesy of Airco.*)

**Figure 19-5** One method of control distortion of the vertical plate.

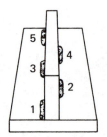

Figure 19-6   Staggered method.

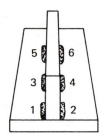

Figure 19-7   Alternating method.

## DISTORTION IN FILLET WELDING

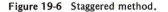

When welding fillet welds, the contraction of the weld metal when cooling causes the vertical plate to become misaligned. This is very common when using a multipass bead on one side of the vertical plate (see Figure 19-4), or when using excessive weld material when making one pass. One way of overcoming the problem is to tip the plate out of alignment before welding (Figure 19-5) and as contraction takes place, the plate will be pulled back to the proper position.

If the plate is accessible on both sides, use a staggered method of welding. That is, weld a short bead on one side of the vertical (Figure 19-6) then a short bead just ahead of the other bead on the reverse side. Continue this procedure the length of the weld. If additional strength is required, go back and fill in between the short beads.

NOTE: Remove the slag.

Another method is to weld a short bead on one side and then a short bead on the reverse side (Figure 19-7). Continue this the length of the plate. One bead will offset the contraction of the bead on the reverse side.

## QUESTIONS

1.   What is a common defect in fillet welding?

2.   What is one of the most common causes of undercutting?

3.   Which electrode may cause more trouble when fillet welding, E6011 or 6013?

4.   What is one cause of distortion of the vertical plate?

5.   How can malalignment of the vertical plate be prevented?

6.   What is another method that can be used to prevent distortion if both sides are accessible?

# OUTSIDE CORNER JOINT WELDS

# chapter 20

The outside corner joint welds are used often in the fabrication of rectangular tanks, machinery, and other structures where the corner must have a good appearance. Excessive penetration, uneven beads, excessive weld metal deposits, or too little weld metal will require additional work to improve the appearance for acceptable quality workmanship; also, tanks may have to be water-tight.

The closed outside corner joint is where the root edges of the plate are touching each other, (Figure 20-1) and when the joint is V-shaped is called an open corner (Figure 20-2). A joint

where one plate covers half of the width of the other plate is called a half joint (Figure 20-3) or half-open joint. This joint is not as strong as the open joint and may require an additional bead on the inside corner of the joint or top plate if bevelled. (Figure 20-4).

A close outside corner joint is weak because of the lack of penetration and proper fusion of the two plates unless there is no strain on the plates involved. If appearance is a factor, grinding would remove some of the metal thus decreasing the strength of the joint. On outside corner joints involving thin plates, a single bead

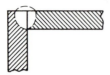

**Figure 20-1**  Closed outside joint.

**Figure 20-2**  Open outside joint.

is generally sufficient for the proper strength. On heavier plates, two or more beads may be required to fill the corner. Welds made on corner joints must be straight and even without melting away too much of the top edges which will cause an uneven appearance. If a tank-like structure is made, it must be liquid tight with no flaws in the weld. With energy conservation, many welders are involved in making wood stoves, especially the airtight type. The

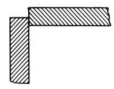

**Figure 20-3**  Half open outside joint.

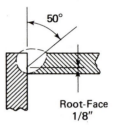

**Figure 20-4**  Beveled or grooved outside corner. Types of outside corner joints.

corners must be neat and smoke tight with no flaws.

## OPEN CORNER WELD

Use plates 1/8″ X 2″ X 8″ (3.17 mm X 50.8 mm X 203 mm) and 1/8″ (3.17 mm) E6013 electrodes (E6011 electrodes are not as good for appearance). Set up plates to form an open V corner and tack weld the joint about every 2″ (50.8 mm) to 5″ (127 mm) to prevent warping of the plates when running the weld. Be sure that the horizontal plate is at a 90° angle from the vertical plate. Run the bead, watching the bead formation and watching for burn through. If holes appear, use a slight weaving motion, or reduce the amperage. Remove the slag and observe the weld for appearance and smoothness. When using 1/4″ (6.35 mm) plate, put down a root bead followed by a second bead using a weaving motion to fill the corner. With two additional plates, weld a half-open corner joint.

## QUESTIONS

1. If appearance is important, what are some types of beads that should be avoided?

2. What are some types of outside corner joints?

3. Which type joint is the weakest?
4. What welds or beads should be made on corner joints?

# HORIZONTAL WELDING

## chapter 21

Many times welding must be done on structures or machinery where it is impossible to position them to weld in a flat position. Welds are made in horizontal, vertical, and overhead positions. Welds in horizontal positions are welding a bead that is parallel to the ground. On beads or welds other than the flat position, the effects of gravity on the molten metal in the puddle will cause the metal to run down on the lower side of the weld. The amount of metal deposited in the weld area will be less thus affecting the strength. Overlapping the lower side of the weld

area with metal without penetration is called cold lapping (see Figure 21-1). If excessive metal runs out of the weld area, it could present a hazard for the welder.

Two important factors that will help to control the molten metal in the puddle are a slight reduction in amperage and the use of a shorter arc than used in a flat position. The use of a weaving motion will help to control the amount of molten metal in the puddle. The electrode angle should be about 15° to 20° in the line of travel with the electrode point upward about

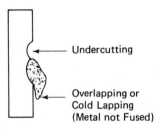

Figure 21-1 Horizontal butt welds.

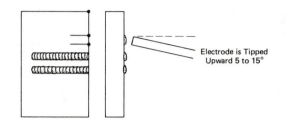

Figure 21-3 Electrode position for horizontal welding.

5° to 15°. If undercutting on the upper part (see Figure 21-1) of the weld takes place, tip the electrode a few more degrees upward.

## HORIZONTAL BEADS

Use plates 1/4″ X 2″ X 6″ (6.35 mm X 50.8 mm X 152.4 mm); plates 1/4″ X 6″ X 6″ (6.35 mm X 152 mm X 152 mm); 1/8″ (3.17 mm) E6011 and E6013 electrodes. Draw a series of lines on a plate 6″ X 6″ (152 mm X 152 mm) about 1/2″ to 3/4″ (12.7 mm to 19 mm) apart and set plate in a vertical position so that the lines are horizontal. Either use a clamp or tack weld to a flat plate for support (see Note and Figure 21-2). First, use the E6011 electrode and set the welder to the recommended current. Run a series of beads or weld across the plate (see Figure 21-3). Remove the slag and check for undercutting and overlapping. Next, use the E6013 electrodes with the proper amperage and run a series of beads or welds across the plate. Check for undercutting, overlapping, and contour. Tack weld a 2″ X 6″ (50.8 mm X 152 mm) plate on the

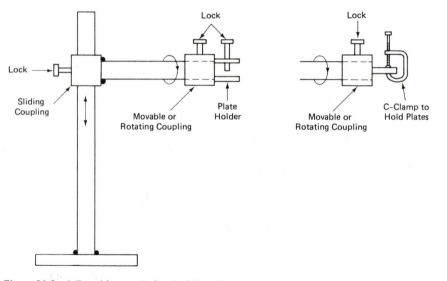

Figure 21-2 Adjustable stand for holding plates for horizontal, vertical and overhead positions.

back of the vertical plate with a 1" (25.4 mm) overlap. Run a single pass lap weld across the joint. Remove the slag and check for undercutting. Repeat the horizontal bead until a noticeable improvement is shown.

NOTE: An adjustable stand can be made out of pipe and a few plates.

## BUTT WELDING IN A HORIZONTAL POSITION

Preparation of plates for butt welding in a horizontal position is similar to butt welding in a flat position. The plates are bevelled to a 30° angle with a 1/16" (1.58 mm) root face on each plate, or they can be ground to a feather edge. When in position, the plates should have about a 1/16" (1.58 mm) gap between the two plates for adequate penetration.

Use steel plates 1/4" to 3/8" X 2" X 6" (6.38 mm to 9.5 mm X 50.8 mm X 152 mm), 1/8" (3.17 mm) E6011 and E6013 electrodes. Bevel one edge of each plate and tack weld together with a 1/16" (1.58 mm) gap between the two plates. Fasten or clamp the plate in a vertical position. Use E6011 electrodes and set the

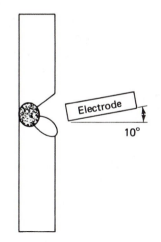

**Figure 21-5** 2nd weld.

welder to the recommended current. Run the first bead deep in the root of the joint maintaining the proper electrode angle (see Figure 21-4). Remove the slag and check for penetration of the two plates. Run the second bead, fusing about 1/3 of the first bead and maintain proper penetration of the lower plate with the electrode tilted downward about 10° (see Figure 21-5). Remove slag. Run the third bead on top of the second and fuse to the upper plate with the electrode tilted upward about

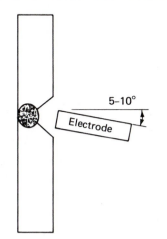

**Figure 21-4** 1st weld.

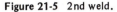

**Figure 21-6** 3rd weld.

10° to 15° (see Figure 21-6). Remove slag. If the groove is not completely filled, add another series of beads. If the groove is nearly filled, add a cover bead using a weaving motion. Remove slag and observe the bead formations. Repeat the procedure but with the top plate bevelled only. If undercutting is present, adjust the electrode angle to overcome the problem. Repeat the procedures using the E6013 electrodes. This type of electrode will present a problem of slag inclusion which can weaken the joint.

## QUESTIONS

1. What is a weld in a horizontal position?

2. What type of force acts on the weld metal in a horizontal position?

3. What is called cold lapping?

4. What are two factors that will help control the weld metal in a horizontal position?

5. If the upper edge of the weld is undercutting, what can be done to correct it?

# VERTICAL
# WELDING

## chapter 22

Welding beads in a vertical position is perhaps the most difficult position because gravity tends to pull the molten metal downward from the weld area. Many welders experience more difficulty when welding up vertical than downward. Vertical welding requires more practice to master than any other position. For maximum strength of a joint in a vertical position, welding upward will deposit more metal than downward. Vertical down welding is suitable for thinner plates because the penetration is not as deep thus preventing burn through holes but still forming adequate welds.

## ELECTRODES

The fast-freeze[1] electrodes are recommended for vertical welding because the weld metal will solidify faster thus preventing the molten steel from running out of the weld area.

Using the E6011 will be easier than the E6013 because of the forceful action of the arc—the E6013 electrode produces a softer arc and the slag is heavier. The angle of the electrode for welding vertical up and down is very

[1] Lincoln

important to help control the molten puddle. When welding downward, the electrode should be pointed upward about 15° to 30° from the vertical plate (see Figure 22-1). The arc should be rather short and the speed fast enough to prevent the molten metal and slag from running ahead of the puddle. When welding upward in a vertical position, the electrode is pointed upward at about 10° to 15° angle (see Figure 22-2). A whipping motion or rocking motion (Figure 22-3) is used to control the molten metal in the puddle. A whipping or rocking motion tips the electrode upward out of the puddle or crater and back to the puddle again. This is simply a slight twist of the wrist and allows the molten metal to solidify.

Repeat this procedure the full length of the vertical bead. During the whipping motion, DO NOT BREAK THE ARC. THE SOLIDIFIED METAL WILL PROVIDE A SHELF FOR ADDITIONAL WELD METAL.

A U-shaped weave and the figure 8 (Figure 22-4) weave type bead is used very often when welding vertical fillet welds. Start with the simple straight bead and then, as this technique is mastered, the weave band will be easier to accomplish.

## WELDING VERTICAL BEADS

### DOWN BEADS

Use 1/4" X 6" X 6" (6.35 mm X 152 mm X 152 mm) steel plates and 1/8" (3.17 mm) E6011 and E6013 electrodes. Set up the plate in a vertical position (Figure 21-1) with a series of vertical lines. Set the welder to the recommended current using the E6011 electrode. Start at the top of the plate and run the bead downward holding the electrode upward at about a 30° angle (Figure 22-1) from the horizontal or 60° from the vertical. Keep the arc short and travel just fast enough to keep the

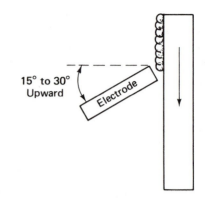

**Figure 22-1** Vertical down welding.

molten metal from getting ahead of the puddle or crater. Remove the slag and check the type of bead formation. Secure three pieces of 1/8" X 2" X 6" (3.17 mm X 50.8 mm X 152 mm) plates. Tack weld two pieces together to form a butt joint with a 1/16" (1.58 mm) gap and secure in a vertical position. Run a vertical down bead with a very slight weaving motion. Remove slag and check for penetration. With the third piece, form a lap joint with the other plate and tack weld. Run a vertical down bead with a slight weaving motion. Next, repeat the same procedure using a E6013 electrode and note the difference between the two electrodes in the amount of slag produced and control of the molten metal.

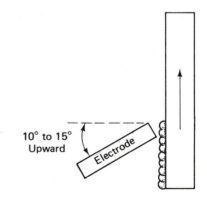

**Figure 22-2** Vertical up welding.

## UP BEADS

Use steel plates 1/4" × 6" × 6" (6.35 mm × 152 mm × 152 mm); and 1/8" (3.17 mm) E6011 and E6013 electrodes. Set up the plate in a vertical position with a series of vertical lines. Set the welder to the recommended current using the E6011 electrode. Start to run a bead from the bottom upward using a whipping or rocking motion (Figures 22-2 and 22-3). Remember, do not break the arc while moving the electrode upward. Remove slag and check the bead for smoothness. Repeat welding upward until the beads become smoother. E6011, with the more forceful arc stream, tends to keep the molten metal from running and produces less slag. The E6013 is more difficult to handle because it has a less forceful arc stream and produces more slag. Use the E6013 electrode next, using the same procedure and note the difference.

In preparation for welding vertical up joints, use different types of weaving motions which will be necessary when welding fillet and lap joints. Figure 8, U-shape, zig-zag, and triangular are some of the weaving motions that could be practiced (Figure 22-4).

## VERTICAL WELDING OF JOINTS

### BUTT JOINT

Use several 1/4" to 3/8" × 2" × 6" (6.35 mm to 9.5 mm × 50.8 mm × 152 mm) plates, E6011 1/8" (3.17 mm) electrodes and E6011 5/32" (3.96 mm) electrodes.

NOTE: Using the E6011 electrode is merely a suggestion because the E6013 and E7014 electrodes can be used in place of the E6011 electrode.

Bevel one edge of each plate to form a 60° groove. Set the welder to the recommended current. Tack weld the plates together with a

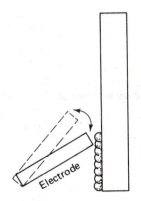

**Figure 22-3** Whipping motion.

1/16" (1.58 mm) gap or root opening. Clamp or fasten it in a vertical position (Figure 21-2). Run a vertical up bead in the root or gap. Remove the slag. Run a second bead with a weaving motion and make sure there is adequate fusion with the first bead and sides of the bevel (see Figure 22-5). Remove slag and check for proper fusion. Run another bead or as many beads as necessary to fill the groove, removing

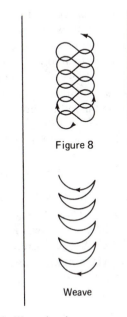

Figure 8

Weave

**Figure 22-4** Weave beads.

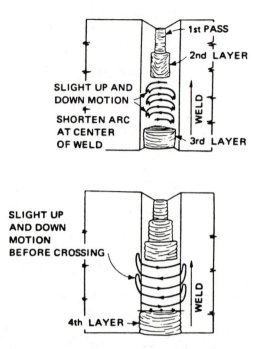

Figure 22-5  Different types of weave beads for butt welding. (*Courtesy of Airco.*)

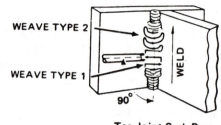

Tee Joint 2nd Pass
Weaving—2 Types

Figure 22-6  Other types of weaving motions. (*Courtesy of Airco.*)

slag after each pass or bead. The final bead should be a cover bead, which would require a wider weaving motion (see Figure 22-5). Try another set of plates to improve your skill in making vertical butt welds. Secure heavier metal, if available, and repeat the same procedure but increase the gap to about 1/8" (3.17 mm). The use of 5/32" (3.96 mm) electrodes could speed up the process of filling the groove.

## FILLET OR T-JOINT WELDING

Use 1/4" to 3/8" X 3" X 6" (6.35 mm to 9.52 mm X 76 mm X 152 mm) and 1/8" (3.17 mm) and 5/32" (3.96 mm) E6011 electrodes. E6013 and E7014 can be used instead of E6011. Tack weld the 2" (50.8 mm) wide plate to the 3" (76 mm) plate to form a T-joint. Set the welder to the recommended current for 1/8"

(3.17 mm) electrodes. Set the plate in a vertical position. Run a vertical up bead in the root area with no weaving motion. Be sure that the weld metal penetrates both plates. Remove slag. One or two more beads can be made to provide adequate strength to the joint. One method is to run a bead fusing the first half of the first bead to one plate. Then fuse the first with the second and the other plate (see Figure 22-6). Remove slag after each bead. The second method is done by using a 5/32" (9.52 mm) electrode with the proper current setting. Run a wide weaving vertical up bead with proper setting penetration of both plates and the root head. For additional strength, another layer of bead can be made with three beads across the joint. Repeat the procedure on the opposite side of the joint.

The use of low-hydrogen electrodes such as the E7018 is increasing in general repair and is also used in welding low and medium carbon steels. It is crack resistant, and elimination of porosity in steels containing sulfur and x-ray quality makes it first choice in some construction and repair work. Some modification in using the low-hydrogen must be made although it can be used in all positions. A very short arc must be maintained in order to secure the proper shielding of the molten metal, therefore, these electrodes should not be whipped when welding vertical. Smaller weaves are better than wide ones.

A slight weaving motion such as triangular or figure 8 produces a satisfactory bead. A slight reduction in amperage will assist in the proper bead formation.

## QUESTIONS

1. Why is welding in the vertical position the most difficult?

2. When welding vertically, which direction is more difficult?

3. Which direction generally is the strongest weld?

4. What kind of vertical welding is generally recommended for thinner plates?

5. Which electrode is easier to use in vertical welding, the E6011 or 6013?

6. When welding vertical down, how should the electrode be held?

7. What is one of the problems that could occur when welding vertical down?

8. When welding vertical up, what motion can be used to control the molten metal?

9. At what angle should the electrode be held when welding vertical up?

10. What are some weaving beads that can be used in vertical up welding?

11. When using low hydrogen electrodes, why are a few modifications of technique necessary?

# OVERHEAD WELDING

## chapter 23

Overhead is a very difficult type of welding because the force of gravity acts against the molten metal in the puddle and normally, the welder is placed in an awkward or uncomfortable position. Overhead welding is dangerous because of sparks or even falling droplets of molten metal falling from the weld area. In many cases, the welder is lying flat on his back when welding overhead on trucks and heavy equipment. It is difficult to make well-shaped beads with the correct penetration because the molten metal tends to sag down. This is why the E6011 electrode or other fast-freeze[1]

---
[1]Lincoln.

electrodes are recommended. The E7018, low-hydrogen electrode, is also quite popular because of the short arc which tends to hold the molten metal in place.

A few saftey precautions should be taken before starting to weld overhead. They are: wear a cap to protect your head and hair; button pockets and make sure the tops of your shoes are covered by pant legs; wear leather sleeves and shoulder covers if available; button shirt collar. If possible, drape the welding over your shoulder to help eliminate arm fatigue if you are standing. When welding overhead, some kind of suitable, adjustable holder for the plates should be available (see Figure 21-2).

The electrode is held at about the same angle as flat welding.

## RUNNING STRAIGHT BEADS

Use several 1/4″ X 6″ X 6″ (6.35 X 152 mm X 152 mm) plates 1/8″ (3.17 mm) E6011 and E6013 electrodes and protective clothing, if available. Draw a series of lines on the plate and clamp it in an overhead position. First try the E6011 electrode and set the welder to the correct current. Start to run a bead using as short an arc gap as possible (see Figure 23-1). Run a series of beads without any weaving motion. If the molten metal starts to drop or sag, reduce the amperage a little. Continue to run the beads in different directions to improve your skill in running overhead beads. When the plate cools, turn the plate over and run a series of beads using the E6013 electrode set to the proper current. Notice the difference in the action of the molten puddle. If the puddle creates a problem of sagging, reduce the amperage. Repeat the procedure but use a weaving motion (see Figure 23-2).

## OVERHEAD WELDING OF JOINTS

### LAP JOINTS

Use several 1/4″ X 2″ X 6″ (6.35 mm X 50.8 mm X 152 mm) plates and 1/8″ (3.17 mm) E6011 and E6013 electrodes. Tack weld two plates together to form a lap joint and clamp them in an overhead position (Figure 21-2). Use the E6011 electrode and set the welder to the proper current. Hold the electrode at a 45° angle (Figure 23-3), the same as welding in a flat position. Run a bead in the root area. Remove the slag. Run the second bead on the wide plate side and remove the slag. Run the third bead fusing the first two

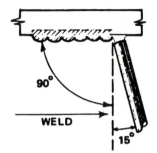

Figure 23-1 Running beads: electrode angle. (*Courtesy of Airco*.)

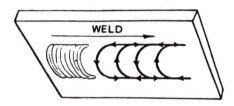

Figure 23-2 Running beads: weaving. (*Courtesy of Airco*.)

beads at the edge of the lap plate or run a fourth bead (see Figures 23-4 a and b). Remove the slag and observe the results. Reverse the plate and run a bead, but use a weaving motion. The deposited metal should go to the top edge of the lap plate and extend about 1/4″ ( 6.35 mm) out on the wide plate. Repeat the pro-

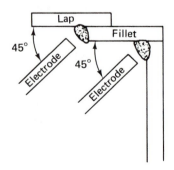

Figure 23-3 Electrode angle for overhead welding.

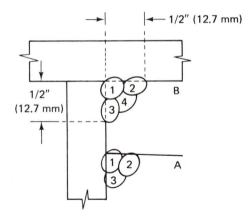

Figure 23-4 Multi-bead welding for lap or inside corner welding in an overhead position.

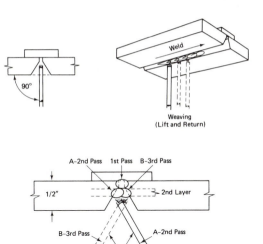

Figure 23-5 Overhead butt joint welding. (*Courtesy of Airco*.)

cedure on another set of plates but change to the E6013 electrode. Beware of slag inclusion when using this type electrode. Remove slag after each bead and observe the results.

## WELDING OF T-JOINTS

Use several 1/4″ × 2″ × 6″ (6.35 mm × 50.8 mm × 153 mm) plates, 1/4″ × 3″ × 6″ (6.35 mm × 76 mm × 152 mm) plates, and 1/8″ (3.17 mm) E6011 and E6013 electrodes. Follow the same procedures used in the lap weld using either a single or multi-bead technique (see Figures 23-3 and 23-4). Remove the slag after each pass. Repeat the procedure using the other type electrode, and watch out for slag inclusion problems of the E6013 electrode.

## WELDING BUTT JOINTS

Use several 1/4″ × 2″ × 6″ (6.35 mm × 50.8 × 152 mm) plates and 1/8″ (3.17 mm) E6011 and E6013 electrodes. Bevel the two edges of the plates to form a 60° angle. Tack weld together leaving about a 1/16″ (1.58 mm) foot opening. Place in an overhead position (see Figure 21-2). With the E6011 electrode and the welder set to the proper current, run a root bead (see Figure 23-5). Remove the slag and run the required bead as in butt welding in a flat position. Change to the E6013 electrode and repeat the above procedure.

## QUESTIONS

1. What are some of the hazards involved when welding overhead?

2. What is the tendency of the molten weld metal in an overhead position?

3. What are some precautions to be taken before welding overhead?

4. What can be done to reduce the sagging of molten metal?

# HARDSURFACING

## chapter 24

Hardsurfacing is a process where a wear-resistant layer of metal is applied to a worn surface to extend the life of machinery or equipment at an economical cost. This unit is merely an introduction to these welding possibilities; it would be impossible to cover it thoroughly, yet keep the unit short. This type of welding technique is not generally used by the average welder but by those involved in working on construction and farm equipment where there is a great need for hardsurfacing.

The type of hardsurfacing and the type of

electrodes used are determined by the service requirements of the parts involved. In many rebuilding jobs, a low cost alloy steel is used and the more expensive alloys are used for severe wear conditions. Some of the types of wear are as follows: metal to metal friction, severe impact, abrasion plus impact, and severe abrasion. Metal to metal friction causes wear from steel parts rolling or sliding against each other with little or no lubrication, such as crawler rollers (see Figure 24-1), drive sprockets, and crane wheels. Severe impact

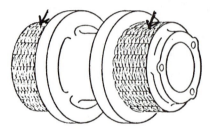

**Figure 24-1** Rebuilding a roller worn by metal to metal contact plus abrasion. (*Courtesy of Marquette Division of Applied Power.*)

causes wear from severe pounding (see Figure 24-2), which tends to squash, gouge, and crack the surface of the metal, such as dipper teeth, buckets, dragline pins, and so on. Abrasion plus impact (see Figures 24-3 and 24-4) causes wear from gritty material accompanied by heavy pounding, which tends to chip or crack as well as grind away the surface, such as dozer blades, shovel tracks, and dipper teeth. Severe abrasion causes wear from gritty materials (Figure 24-5) like sand that grinds or erodes

**Figure 24-3** An example of worn bucket teeth caused by abrasion and impact. (*Courtesy of Lincoln Electric.*)

the surface. This kind of abrasive wear is often accompanied by heavy compression or medium impact. It is found in grader blades, bucket lips, screw conveyers, and so on.

For the best results, know what type of wear condition exists on the part or parts involved. Then consult the manufacturer's electrode guide for the proper electrode and procedure to follow. On some parts subjected to severe wear, the entire surface or the wear area may be covered with hardsurfacing. The area to be

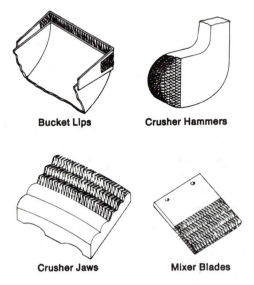

Bucket Lips      Crusher Hammers

Crusher Jaws      Mixer Blades

**Figure 24-2** Rebuilding parts that will be subject to severe impact. (*Courtesy of Hobart Brothers.*)

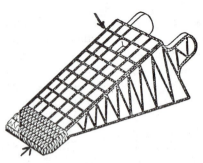

**Figure 24-4** A method of rebuilding bucket teeth. Note the different patterns. (*Courtesy of Marquette Division of Applied Power.*)

hardsurfaced must be cleaned of rust, grease, and other foreign materials by grinding to a

**Figure 24-5** Rebuilding the edges of an auger with hard surfacing subjected to severe abrasion. (*Courtesy of Lincoln Electric*.)

bright surface.

When welding, avoid using too much amperage as it will cause excessive mixing of the base metal and the hardsurfacing alloy, which will cause the hardsurfacing deposits to become too soft. Use a longer arc and a low amperage just high enough to fuse the beads together. On some hardsurfacing electrodes a shorter arc and a wide weave pattern is recommended.

Types or patterns of deposits are not usually important. In some cases, the best pattern is usually the one most economical to apply (Figure 24-6). A pattern with openings between the beads is practical when the openings fill with abrasive materials, thus protecting the base metals. Welding beads across the flow of materials helps pull the materials through the rolls—such as crusher rollers. Beads parallel with the flow of materials will support the heavy material while offering little resistance to the flow. A diamond-like pattern is often used because it is self-cleaning.

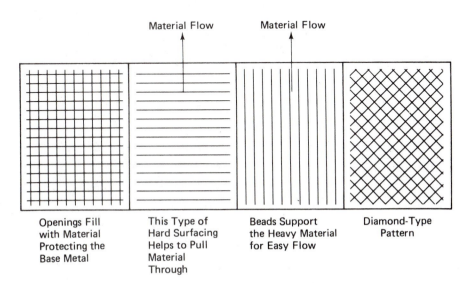

Material Flow   Material Flow

Openings Fill with Material Protecting the Base Metal

This Type of Hard Surfacing Helps to Pull Material Through

Beads Support the Heavy Material for Easy Flow

Diamond-Type Pattern

**Figure 24-6** Some types of hard surfacing bead formations.

## QUESTIONS

1. What is hard surfacing?
2. What are some of the types of wear?
3. What determines the electrode to be used?
4. What is one precaution to be followed when hard surfacing?
5. Generally, how should the surface be prepared?

# CARBON ARC WELDING

## chapter 25

Carbon arc welding is another form of electric arc welding where electricity is used to supply the necessary heat for fusion. One or two carbon electrodes are used. This kind of welding can be used for brazing, steel welding, heating metal to bend it, loosening rusted bolts, hard-surfacing, heat treating metal, and even soldering. When using a single carbon electrode, the heat is produced by an electric arc (or resistance) created by touching the carbon electrode to the work piece. In the two- or twin-carbon torches (see Figure 25-1), the heat is produced by an electric arc between the two carbon electrodes. The electrodes are generally copper coated with soft carbon centers of a pure graphite or baked carbon, in sizes from 1/8"

(3.17 mm) to 1/2" (12.7 mm) in diameter. The size of the carbon electrodes, the length of the arc, and the amperage determines the amount of heat produced. The carbon arc torch can be used on either AC or DC welding machines, but it is easier to use the twin-carbon torch on the AC machine. If DC machines are used, the positive carbon electrode should be one size larger than the negative carbon electrode. When using a single carbon electrode on a DC machine, the current used is straight polarity.

*CAUTION: The rays produced by the carbon arc are just as dangerous as those produced by the shielded electrode. The regular welding helmet and proper clothing should be worn.*

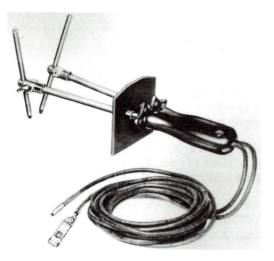

**Figure 25-1** Twin carbon arc torch. (*Courtesy of Lincoln Electric.*)

*When adjusting the carbons, turn off the welder to prevent flash burn to the eyes in case the carbons accidentally touch each other.*

## PROCEDURE

With the welder turned OFF, connect one lead of the carbon arc torch to the ground clamp and the other to the electrode holder.

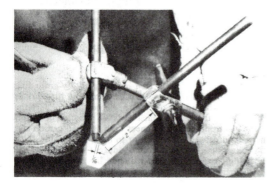

**Figure 25-2** Adjusting the carbon electrodes. (*Courtesy of Lincoln Electric.*)

NOTE: On some carbon arc torches, the leads are marked.

In this procedure, the work piece is not grounded. Adjust the carbons so they extend about 2″ to 3″ (50.8 mm to 76.2 mm) below the clamps with the ends of the carbon bevelled (see Figure 25-2). The carbons should have about a 1/16″ (1.58 mm) gap in the closed position. Set the welder to the correct amperage in relation to the size carbons used and the thickness of the metal to be welded (Figure 25-3).

Check with charts, if available. Separate the carbons and turn the welder on. Lower the weld helmet over your face and start the arc by bringing the carbons together. Then draw them apart until the flame is stable and quiet. The gap between the two carbons is usually about a 1/4″ (6.35 mm). The amount of heat on the work depends on the distance of the carbon arc to the work. With a 5/32″ (3.96 mm) to 1/8″ (3.17 mm) mild steel gas welding rod in the opposite hand, heat one end of the joint to be welded by moving the torch 2″ or 3″ (50.8 mm to 76.2 mm) back and forth along the joint (see Figure 25-4). Keep the torch parallel with the joint or (Figure 25-5) one carbon on each side of the joint. Heat the work piece until a puddle starts to form in the joint, then add the filler rod. Move the torch slowly along the joint and add the necessary filler rod—be sure that both pieces are fused together. Extinguish the flame by moving the carbons apart. Turn off

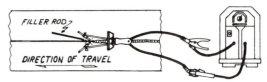

FILLER ROD

DIRECTION OF TRAVEL

**Figure 25-3** Positioning electrodes. (*Courtesy of Lincoln Electric.*)

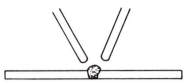

**Figure 25-4** Carbon arc guide.

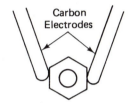

**Figure 25-5** Keep torch parallel with the joint.

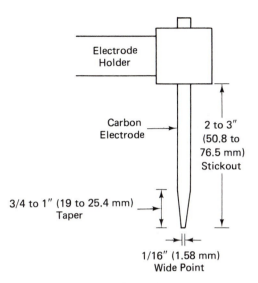

**Figure 25-6** Loosening nuts by heating them with the carbon electrodes.

the welder and place the torch in a safe place until the carbons cool. To loosen rusted nuts, turn the machine on (Figure 25-6), then touch the opposite sides of the nut with the carbons for a few seconds. The area of the nut touching the carbon should turn red; the nut can then be removed with a suitable wrench.

## SINGLE CARBON OPERATION USING DC WELDER

Some electrode holders will not accommodate the carbons so a special holder is available for use with one of the twin-arc holders. The carbons used in the single-arc torch should be ground to a taper about 3/4" to 1" (19 mm to 25.4 mm) long with a point 1/16" (1.58 mm) wide. The length of stick out from the holder varies with different types of carbon holders, about 2" to 3" (50.8 mm to 76.5 mm). The fine point or thin carbons are used for thin metal.

With the welder turned off, insert the tapered carbon in the holder and adjust it to the proper stickout. Set the welder to proper amperage in relation to the diameter of the carbon, for example, for 1/4" (6.35 mm) carbon, set the welder from 50 to 80 amps DCSP. Attach the ground clamp to the work piece. Turn the welder on and lower your welding helmet. Bring the carbon in contact with the work piece to start the arc, then withdraw it to the proper arc gap, about 1/4" to 3/8" (6.35 mm to 9.52 mm), depending on the thickness of the metal. If the gap is too wide, the flame will go out. If the carbon burns red more than 1½" (38 mm) from the end, the amperage is too high. Use a mild steel filler rod if it is compatible with the work piece metal or use another kind of filler rod for the type of metal being welded. Hold the filler rod in your opposite hand, heat the joint until a puddle of molten metal forms and then add the filler rod. Keep the torch moving along the joint and only add as much filler rod as needed. Raise the carbon to extinguish the flame, then turn the welder off. Place the carbon in a safe place until the carbon cools.

## SOLDERING WITH A SINGLE-CARBON ARC USING EITHER AC OR DCSP

Soldering can also be done with a carbon that is pointed. It can be used on either AC or DCSP machines. When soldering, the amperage used is about 20 to 30 amps. The leads are the same as in single-carbon welding.

Properly clean the area to be soldered, then coat the area with a solder paste or flux. With the welder turned off, adjust the carbon and properly taper it to a fine point. Connect leads as previously described in single-arc welding. A welding helmet is not required for this procedure because there will be no arc flame. Turn the machine on and touch the carbon to the area to be soldered. The electrical resistance between the carbon and the work piece will supply the necessary heat. Apply the solder with the other hand and move the carbon along the area being soldered. Remove the carbon from the metal as soon as the soldering is finished and turn off the machine. Place the carbon in a safe place until it cools.

**Table 25-3**

| THICKNESS OF BASE METAL | APPROXIMATE CURRENT SETTING | CARBON DIAMETER |
|---|---|---|
| 1/32" .794 mm | 30–50 amps | 1/4" 6.35 mm |
| 1/16" 1.58 mm | 50–60 amps | 1/4" 6.35 mm |
| 1/8" 3.17 mm | 70–80 amps | 5/16" 7.93 mm |
| 1/4" 6.35 mm | 90–100 amps | 3/8" 9.52 mm |
| over 1/4" 6.35 mm | 110–125 amps | 3/8" 9.52 mm |

## QUESTIONS

1. What produces the heat when using a single and twin carbon?
2. What determines the amount of heat produced?
3. How dangerous are the rays from the carbon arc?
4. How far should the carbon be extended below the clamp?
5. What is the recommended gap between the carbon in a closed position?
6. How far apart should the carbon be to obtain a stable and quiet arc?
7. How should the carbons be placed along the joint?
8. How is the flame or arc extinguished when using twin carbons?
9. How long should the taper be in a single carbon?
10. What type joint of the single carbon should be used on thin metal?

11. What is an indication that excessive amperage is being used in the single carbon?

12. When soldering with a single carbon, what is the approximate amperage?

13. What causes the necessary heat?

14. Is it necessary to use a welding helmet when soldering?

# WELDING ALUMINUM WITH METALLIC ARC

## chapter 26

Aluminum, a lightweight non-ferrous metal, with its high thermal and electrical conductivity and corrosion resistance, plays an important part in the manufacture of many items. They range from coffee pots, cooking utensils, automobile hoods, and truck bodies to railroad cars and aluminum foil for food wrapping. Commercially pure aluminum in a cast or annealed state has about one-fifth the tensile strength of structural steel. In some cases, the aluminum is as strong as steel. One of the big advantages of aluminum is the high strength to weight ratio. In its pure state, aluminum is very ductile and can be easily shaped and formed. Automobile manufacturers are looking toward aluminum because of its light weight. The

future is bright for aluminum and welding of aluminum will become more important. At the present, many welders shy away from anything made of aluminum.

## COMPOSITION OF ALUMINUM PRODUCTS

Besides pure aluminum in sheet, tubing, or other structural shapes, aluminum casting and wrought alloys contain other elements such as silicon, zinc, nickel, copper, manganese, and others to give it strength, or other desirable properties. Aluminum is frequently alloyed with manganese and silicon to make it corrosion resistant. Cast aluminum or castings made of aluminum are frequently encountered in the repair field. They include transmissions, housings, snowmobile gear cases, and engine parts. Cast aluminum generally contains a higher percentage of alloying elements than wrought alloys. General purpose aluminum castings contain copper, silicon, and a small percentage of iron with silicon used as a hardener. Most of these castings can be welded without many problems.

The type of electrode used most frequently contains 95% aluminum and 5% silicon with a heavy flux coating. The coating dissolves the oxide film in the weld area and stabilizes the arc. The slag, like steel welding, prevents oxidation of the weld metal while cooling. Aluminum electrodes give the best results when used with DC reverse polarity. Welding aluminum sheets or castings that are less than 1/8" (3.17 mm) thick may create a problem of arc control at reduced amperage. If the amperage is too high, the base metal will melt away.

Due to the fact that it is easy to get good penetration, little or no joint preparation is necessary when welding aluminum up to 1/4" (6.35 mm) thick. Even up to 3/8" (9.52 mm) thick, no preparation is necessary if both sides are welded. When repairing casings or housing that contain oil or gear lubricant, it is advisable to V-groove along the fracture. If the oil or lubricant remains in the fracture at the time of welding, it could cause the weld to be defective because of the porosity and contamination.

The use of backup plates is advised whenever possible to prevent burn through. Because of the high heat conductivity, preheating the casting or metal from 250°F to 400°F (121°C to 208°C) will improve the weld puddle and will help to minimize distortion. The scratch method of starting the arc is recommended to avoid having the electrode stick to the metal. Because the electrode melts away very fast, keeping a short arc when welding may be difficult. A long arc will produce excessive splatter and the arc will be difficult to control. Keep the electrode in a near vertical position for better arc control. Use an arc about 1/8" (3.17 mm). Move the electrode in a straight line and avoid any weaving motion. After welding, be sure to remove the slag; if not removed, the slag will start to corrode the surface. Follow the manufacturer's recommendation for slag removal, then wash the area with warm water.

Use scrap sheet aluminum 3/16" to 1/4" (4.76 mm to 6.35 mm) thick, 3" X 6" (76 mm X 152 mm) pieces, 1/8" (3.17 mm) coated aluminum electrodes, and copper backup plates, if available. Remove all traces of grease, oil, or other foreign material. Mechanically remove oxides along the weld area by wire brushing or scrubbing with steel wool. Position the two pieces of aluminum to form a butt joint with a 1/16" (1.58 mm) space between them. Use a backup plate if available. Set the welder or DCRP and the recommended amperage of the electrode. Scratch start the electrode and follow straight along the joint, without weaving; keep the electrode in a vertical position with a short arc. After cooling, remove slag by chipping and wire brushing. Observe the type of bead formed. Repeat the procedure by adding another piece. Because of the fast burn off rate of the electrode, it takes a lot of practice in order to obtain a fairly smooth bead. Remove slag after cooling and wash the samples in warm water if they are to be saved.

## WELDING CASTINGS OR GEAR HOUSINGS

Use broken or cracked transmission housings or other castings (these are readily available from salvage yards) and 1/8" (3.17 mm) coated aluminum electrodes, E-A1-43. Remove all traces of oil, grease, dirt, or other foreign material from the area to be welded. Mechanically remove oxides or paint from weld area. V-groove along fracture lines or bevel pieces of broken sections. If pieces are broken out, clamp in position. Set the welder to the amperage recommended by the electrode manufacturer. Preheat the casting from 400°F to about 800°F with an oxy-acetylene torch. Be careful not to overheat the casting or to hold the torch too close. Scratch the electrode, holding the electrode vertically and use a short arc weld along the fracture lines. When welding broken pieces, tack the weld and check for proper alignment. After cooling, remove slag and check for fusion and holes in beads.

Although aluminum can be welded successfully with the stick electrode, the rapid burning of the electrode and the limitation in thickness of the material to be welded makes aluminum welding by stick electrode difficult. Other methods, such as MIG and TIG welding of aluminum, are easier and more controllable.

## QUESTIONS

1. What is one of the big advantages of aluminum?

2. What are some of the elements that are added to aluminum to give some desirable properties?

3. What is frequently added to aluminum to make it corrosion resistant?

4. What type joint preparation is necessary for aluminum up to $\frac{1}{4}$" (6.35mm) thick?

5. Why is it necessary to clean and V-groove housings that contain oils or lubricants?

6. Why is it advisable to preheat the casting before welding?

7. What arc length should be used?

8. Why is it important to remove the slag?

9. What method is best for starting arc?

10. Why are backup plates recommended wherever possible?

# PIPE WELDING

## chapter 27

Welding has become the accepted method of joining pipes together, especially in the construction of pipelines used for all transportation of gas, oil, and other substances. On small diameter pipes, welding eliminates costly threading operations and the use of couplings. Pipes are used in construction replacing beams, channels, and other structural shapes. Welding of pipe is also used in joining low-pressure pipes in refrigeration, heating, and air-conditioning systems. Although pipes are jointed by other means, such as automatic or semi-automatic MIG, TIG, and, in smaller pipes, oxy-acetylene welding, the stick electrodes or the shielded arc is still being used in many areas.

Pipe welding is in a field by itself because of the types of liquids or gases that are transported through pipes. Because of public safety, environmental restrictions, and the health hazards of some liquids, pipe welders must pass certain tests to be certified to weld on these pipelines. The tests will vary in different states and with local construction codes.

## PIPE WELDING POSITIONS

In the previous procedures, the plane or surface of the plate to be welded always remained constant whether it involved flat, vertical, or overhead welding (see Figure 27-1). In pipe

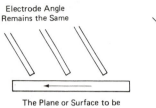

Electrode Angle
Remains the Same

The Plane or Surface to be
Welded Remains the Same

In Pipe Welding the
Electrode Must Follow
the Surface Which is
Constantly Changing

**Figure 27-1** Difference between flat and circular (pipe) welding.

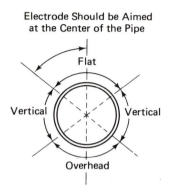

Electrode Should be Aimed
at the Center of the Pipe

Flat

Vertical

Vertical

Overhead

**Figure 27-3** Approximate welding positions when welding pipe in a fixed or stationary position.

welding, the plane is constantly changing as the welder runs a bead around the pipe or in a circle. But the electrode must be held at a constant angle to the pipe.

Pipe welding normally involves three positions. They are a horizontal fixed position, a horizontal rolled position, and a vertical position. The horizontal rolled position is the easiest position because the pipe is rolled and the welder is welding in a flat (see Figure 27-2) or near flat position. In the horizontal fixed position, the position of the weld is constantly changing from flat (see Figure 27-3) to vertical to overhead. The vertical position is welding in a horizontal plane (see Figure 27-4) with the control of the molten metal creating the problem. The use of the fast-freeze electrode helps to control the puddle in the different positions. In some codes or specifications, other types of electrodes may be required.

To practice pipe welding, a roller-type device to turn the pipe is very helpful, but pieces of angle iron (see Figure 27-5) can be used to hold the pipe in position. In some pipe joining operations, a backup ring with small knobs is used to keep the pipe properly aligned (see Figure 27-6). The knobs for the proper root opening and the ring will prevent the molten metal from sagging. Special jig and clamps are also used to keep the pipe in proper alignment and maintain the proper root opening.

After the pipe is aligned and you are preparing to weld the pipe, it is best to tack weld in three or four places prior to welding a solid

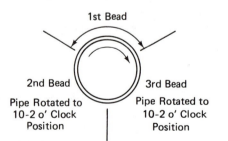

1st Bead

2nd Bead

Pipe Rotated to
10-2 o' Clock
Position

3rd Bead

Pipe Rotated to
10-2 o' Clock
Position

**Figure 27-2** Movable pipe: horizontal rolled position 1st bead, about 10 o'clock to 2 o'clock.

**Figure 27-4** Vertical fixed position, welding in a horizontal position required. Three positions of pipe welding.

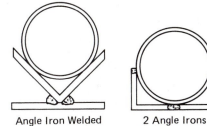

Angle Iron Welded      2 Angle Irons
to a Flat Plate        Welded Together

**Figure 27-5** Some devices that can be used to hold pipe in position or keep from turning.

bead. These positions and welding positions are often referred to as clock positions. A four-tack weld procedure would be mentioned as 12, 3, 6, 9 o'clock tack welds. Welding in a rolled position would be running a bead from a 10 o'clock to a 2 o'clock position (see Figure 27-2).

## PROCEDURE

Joint preparation is very similar to the weld butt joint on plate steel. On large diameter pipes, the edges of the pipe should be bevelled to a 30° angle to form a 60° angle joint with a 1/16" (1.58 mm) root face, if possible. The pipe can be bevelled with a torch and touched up with a grinder, especially when forming the root face. Proper penetration and fusion are very

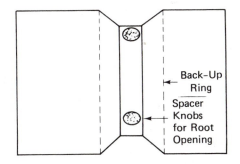

Back-Up Ring

Spacer Knobs for Root Opening

**Figure 27-6** A sketch of a back-up ring used for alignment and the proper root opening.

important for strength and to make sure the piece is leakproof. Metal sagging through the joint, commonly referred to as icicles, must be avoided. When running practice beads on pipe, it is very difficult to run a bead around the surface of the pipe and have the ends of the bead meet. This can be done by wrapping a piece of heavy paper or thin sheetmetal around the pipe and marking the line with soapstone. This line is also very helpful in marking pipe to be cut with a torch.

### PRACTICE

Use steel pipe 3" × 6" (76 mm × 152 mm) in diameter and about 12" (304 mm) long. On standard pipe, the pipe size is indicated on the inside diameter. Use 5/32" (3.96 mm) and/or a 3/16" (4.76 mm) electrode, E6011 and E6010. You can use a pipe roller or angle iron for pipe support. Mark a series of circular lines on the pipe as previously discussed. Support the pipe from rolling. Set the welder to the correct current for the size of electrode used. If E6010 is used, set welder at DCRP. Start the bead with the electrode aimed at the center of the pipe at all times. Use a 10 o'clock position and run bead to a 2 o'clock position. Rotate the pipe by one-third and restart bead at 2 o'clock going to 6 o'clock. Rotate the pipe one-third and finish the bead. Remove slag before continuing the bead. Repeat the procedure on the next line, until a satisfactory weld is produced. Also, with practice, the starting points should be hard to detect. For left-handed welders, reverse the procedure by going from 2 o'clock back to 10 o'clock and so on until the weld is complete.

### PIPE WELDING—CLOSED BUTT JOINT, ROLL POSITION

Use steel pipe 3" to 6" (76 mm to 152 mm) in diameter, several pieces 3" (76 mm) long, 5/32" (3.96 mm) E6011 or E6010 electrodes, and a pipe roller or angle iron for support.

Position the pipe in the support, then tack weld at three or four equally distant spots. Set the welder to the proper current. Run beads similar to the practice with one exception: proper and uniform penetration must be maintained without burn through or icicles on the inside surface of the pipe.

## WELDING PIPE—OPEN BUTT JOINT, ROLLED POSITION

Repeat the procedure as in the closed butt, but space the pipes about 1/16" (1.58 mm) apart. After welding, check for uniform penetration without burn through.

## PIPE WELDING—BEVELLED JOINT, ROLLED POSITION

Use 3" to 6" (76 mm to 152 mm) diameter pipe, 3" (76 mm) long, 1/8" (3.16 mm) and 5/32" (3.96 mm) E6011 or E6010 electrodes, and a pipe roller or angle iron for support. Bevel the end of two pieces of pipe to a 30° angle leaving about 1/16" (1.58 mm) root face. Set the welder to the proper position for a 1/8" (3.17 mm) electrode. Tack weld in three or four equally spaced locations. Run a root bead with proper penetration and without burn through. Remove the slag and check for penetration. Use the 5/32" (3.96 mm) electrode with the proper amperage. Run the bead with a slight weaving motion; keep the electrode aimed at the center of the pipe at all times.

## PIPE WELDING BUTT AND BEVELLED JOINTS—FIXED POSITION

Welding pipe in a fixed position can be done by using the same type of holder or positioner used in overhead welding, especially if it is adjustable and can be swung out of any position. The bead can be started at 12 o'clock and continued around the complete circumference of the pipe—involving vertical, overhead, and horizontal welding. A downhill method can be used by welding down vertical from 12 o'clock to 6 and from 12 o'clock backward to 6. Use the same precautions as mentioned in vertical down welding as to puddle control. Practice this procedure until satisfactory welds are made.

Another practice procedure that could be used to simulate field repairing is to fasten the pipe within about 2 ft. (608 mm) from the floor and then practice welding in that position. The bottom of the pipe would have to be welded by the welder lying on the floor, welding overhead.

## QUESTIONS

1.  Compare the major difference between flat and pipe welding?

2.  What are the three positions in pipe welding?

3.  Which one is the easiest to weld?

4.  What position is the most difficult?

5.  What is sometimes used to keep pipe in alignment with the proper root opening?

6.  Generally, how many tacks should be made before welding?

7.  What is meant by "icicles"?

8.  Besides a strong welded joint, what is another thing that is important when pipe welding?

# CAST IRON

## chapter 28

The name "cast iron" covers a wide range of different products. Some products made of cast iron are very brittle and weak; these include counterweights or structures where strength and appearance are not important. Other products made of cast iron are strong and can withstand heat and pressure; these include pump housing, engine blocks and heads, and so on.

## TYPES OF CAST IRON

There are five types of cast iron that contain a high percentage of carbon. They are: gray, white, alloy, modular, and malleable. The content of carbon ranges from about 1/8 to about 4%. Silicon content ranges from 0.85% to 2.85% with the silicon content in white cast being much lower.

In gray cast, the metal is permitted to cool slowly, and as a result, the carbon separates in the form of flakes called free carbon. This is the reason why gray cast is very brittle. When fractured, the dark porous structure is easily identified with gray iron. Gray iron can be welded without too much difficulty.

In white cast, on the other hand, the carbon is combined resulting from the rapid cooling which leaves the metal very hard and difficult to machine. When fractured, the fracture will appear as a fine silvery white formation. However, abrasion and wear resistance are essential, and therefore it is practically impossible to weld or repair.

Alloy casts are those where different alloying elements are added to the cast iron to improve the machinability, the tensile strength, resistance to corrosion, or other necessary qualities. Pre-heating and post-heating are very essential when welding alloy cast so as not to destroy the alloying elements. Nodular cast has the corrosion resistance of alloy cast, greater tensile strength than gray cast, and the ductility of malleable cast. The annealing process and the addition of magnesium causes the carbon or graphite to form in the shape of nodulars (spheres). This process improves the mechanical properties of the cast. Malleable cast is white cast that has been annealed, leaving the metal soft and making the metal tougher. When fractured, the center will be dark with the outer edges a lighter color. If malleable cast is welded, high heat must not be used because it will turn the malleable cast back into white cast.

Another type of cast iron is called chilled iron. It is similar to gray cast, but it has a lower silicon content. The casting is chilled, which creates a surface of extremely hard white iron, free from graphite carbon. The center of the casting will have the typical gray iron appearance. The outer area will be wear resistant of white cast; the center will be gray, which is less brittle.

## PREPARATION

When preparing to weld cast iron, the outer surface of the structure must be ground off (often called the skin) because it contains impurities from the mold and other impurities that rose to the surface before the metal hardened. On thin casting, a V-groove should be made along the fracture line, about one-half the thickness of the cast. This can be done

**Figure 28-1** Different pieces of cast prepared for welding. (*Courtesy of Linde Division of Union Carbide.*)

with a grinder, a diamond point chisel, or an air impact chisel, if available (see Figure 28-1). On casting that is heavier, a V-groove (Figure 28-2a) must be made with about a 1/16" (1.58 mm) root face with the groove at approximately a 60° angle, similar to the angle used for steel welding. If both sides are accessible, a double V-groove (Figure 28-2b) can be used with a 1/16" (1.58 mm) root face in the middle. Sometimes the fractures increase in length when welding. This can be controlled in many cases by drilling a hole (Figure 28-3) a short distance from each end of the crack. Be sure to remove rust, grease, dirt, or other foreign material that may be trapped in the weld and weaken it. Also, be sure to remove the skin or surface from the weld area.

When checking a casting for small fractures, rub a small amount of kerosene over the area and wipe dry. Then wipe some white chalk over

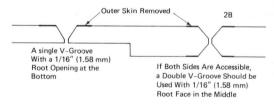

Outer Skin Removed    2B

A single V-Groove With a 1/16" (1.58 mm) Root Opening at the Bottom

If Both Sides Are Accessible, a Double V-Groove Should be Used With 1/16" (1.58 mm) Root Face in the Middle

**Figure 28-2** Some different types of joint preparation for cast iron welding.

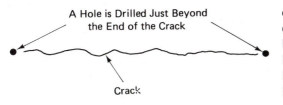

A Hole is Drilled Just Beyond the End of the Crack

Crack

**Figure 28-3** A crack in a cast iron casting.

the area. The kerosene in the cracks will turn the chalk gray or produce a wet line. Pin prick along the line so it will not be lost when V-grooving the fractures. Keeping the casting as cool as possible when welding is very important. If not, additional cracks may form. In the case of malleable cast, the cast will change to white cast and become very hard. Preheating the casting will help to eliminate cracking if the casting is preheated evenly and never beyond a dull red. The casting must be kept at a temperature between 500°F and 1100°F (260°C to 593°C) and allowed to cool slowly after welding. Covering the casting (if not too large) with dry sand or asbestos will help slow the cooling. Never immerse a heated casting in water as it is done with mild steel.

For large casting and many other types of casting where pre-heating is difficult, a series of short beads, about 1" to 2" (25.4 mm to 50.8 mm) long are made, letting the casting

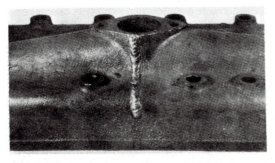

**Figure 28-4** A shielded metal-arc weld made on a cast iron cylinder head. (*Courtesy of Lincoln Electric.*)

cool down between each bead. As each bead cools, peen lightly with a hammer on the new bead, not the whole casting. The peening of the bead helps to relieve the stresses around the bead area. Another technique that can be used is the back step method similar to that in steel welding, however, with cast iron each bead must be allowed to cool.

## ELECTRODES

There are two main types of electrodes used for cast iron welding. They are machinable—nearly pure nickel which leaves a soft ductile deposit that can be machined. The other electrode is non-machinable, made of mild steel core with a heavy coating that melts at a low temperature. The deposits are very hard and cannot be machined. It is ideal for repairing transmission cases, engine blocks (Figure 28-4), and pump parts because the electrode produces a waterproof and tight bead.

When welding cast iron, keep the amperage set as low as possible yet maintain the proper fusion. If the amperage is too high, the weld will boil and cut into the casting too far. If the amperage is too low, the weld metal will not flow but pile up and the arc will be too difficult to maintain. The arc should be slightly longer than when using mild steel electrodes.

## PROCEDURE

For practice welding cast iron, obtain scrap transmission cases from the local salvage yard and crack them with a hammer. Use 1/8" (3.17 mm) machinable and non-machinable electrodes, scrap castings, a diamond point chisel and hammer, safety glasses, and a grinder.

V-groove the fractures with a diamond point chisel and grind off the surface of the casting along the fractures. Wear safety glasses. Wipe with solvents to remove grease, oil, or other foreign matter. First, try the machinable type electrode. Set the welder to the recommended electrode current. Run a 1" to 2" (25.4 mm to 50.8 mm) bead at one end of the fracture, remove the slag, and peen lightly with a hammer on the bead. Run a 1" to 2" (25.4 mm to 50.8 mm) head at the opposite end of the fracture, remove the slag, and peen lightly. Let the casting cool a little before proceeding. Run a bead the same length in the middle of the fracture, remove the slag, and peen lightly. Repeat this procedure until the fracture is completely welded. Always run the bead in the coolest spot of the casting.

Another technique is to try the back step method as described in the steel butt weld section. Start 1" to 2" (25.4 mm to 50.8 mm) from the end of the fracture and weld to the end. After cleaning and peening and sufficient cooling, advance another 1" to 2" (25.4 mm to 50.8 mm) and weld a bead back to the first bead and so on until the fracture is welded. This involves a little more time because the casting must not get too hot. Repeat the above procedure using the non-machinable electrodes and set machine to the recommended current.

Many types of casting or machinery parts can be repaired with success, but it takes time. In working with cast iron, haste makes waste. In some cases where a gear box or like structure is broken and pieces missing, pieces of mild steel can be inserted and welded to the cast iron.

## QUESTIONS

1.  What are the five types of cast iron?
2.  What kind of cast iron is very brittle?
3.  Which type is very hard and difficult to machine?
4.  What is the range of carbon in cast iron?
5.  Preheating and post-heating are essential for what kind of cast iron?
6.  What kind of cast iron is corrosion resistant, very strong, and ductile?
7.  Why is it necessary to remove the skin of the cast iron before welding?
8.  How should the metal be prepared before welding?
9.  In some cases, how can further cracking of the casting be prevented?
10. How do you weld large castings too difficult to preheat?
11. What are the two most common types of cast iron electrodes?
12. What is the danger of using high amperage when welding cast iron?
13. What is the purpose of peening?

# CUTTING WITH METALLIC ELECTRODES

## chapter 29

Many times, when repairing equipment or other articles, metal or bolts must be cut when an oxy-acetylene torch is not available, especially in field work. Heavy and thin metal may be cut with a different procedure using the same type of electrode as used for welding. There are some electrodes made especially for cutting metal. The cutting involves the intense heat of the electric arc to melt the metal.

## USE OF THE ELECTRODE

In cutting, the amperage is set very high in relation to the size of the electrode used and the thickness of metal to be cut. For example,

cutting with an 1/8″ (3.17 mm) electrode and metal thickness ranging from about 1/8″ (3.17 mm) to 1/2″ (12.7 mm) thick, the amperage setting would range from about 140 to about 225 amps depending on the type of machine used. The force of the arc pushes the molten metal away; some of this force is generated by a jet action from the coating of the electrode at high temperature. Steel, cast iron, and some non-ferrous metal can be cut with the arc. Because of the intense heat generated by the higher amperage, more protection is needed, especially for the hands. Also, avoid the fumes when cutting galvanized metal, brass and bronze material, or material covered with paint or exposed to chemicals.

Be sure to cut in a well-ventilated area, or if possible, position yourself with the wind at your back so the fumes and gases are blown away, not toward the operator. Check the area for flammable material because the sparks will travel a lot further than when welding.

Many welders prefer the E6011 electrode for cutting because of the jet action created by the coating. This jet action can be increased by soaking the electrodes for about five minutes in water. This moisture slows down the burning of the coating and produces a deeper cavity on the end of the electrode, which creates a stronger jet action.

*CAUTION: Do not use these water-soaked electrodes for welding because the hydrogen in the water will weaken the weld. Keep them in a separate area if used frequently or soak as needed.*

A carbon used in carbon arc welding (see carbon arc section) can be used for cutting but it does not have the force created by the metallic electrode.

An air-carbon arc cutting attachment is also used but this involves the use of a special attachment to hold the carbons plus a compressed air supply. This type operation requires a higher capacity welder.

## PROCEDURE

Use scrap metal 1/8″ to 1/4″ (3.17 mm to 6.35 mm) thick and 1/8″ (3.17 mm) E6011 electrodes. It is also possible to use E6013 electrodes. Position the metal so that the cutting area is off the edge of the bench. Check the area for flammable material and use sand on the floor to catch the molten material. Put the electrode in the holder and set the amperage from 150 to 175. If using an E6013, set the amper-

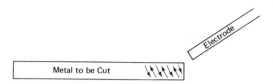
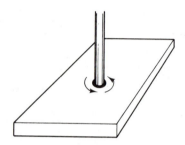

**Figure 29-1** Cutting with electrodes. Keep electrode at a low angle and move electrode upward, then downward to force out the molten metal.

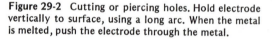

**Figure 29-2** Cutting or piercing holes. Hold electrode vertically to surface, using a long arc. When the metal is melted, push the electrode through the metal.

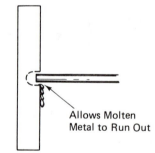

Allows Molten Metal to Run Out

**Figure 29-3** For heavier plates, use this position for piercing holes. Allows molten metal to run out.

age a little higher. Strike the arc on the edge of the plate and hold the arc long until the edge starts to melt. Move the electrode close and move downward to force the molten metal from the plate. Keep moving the electrode up and down across the thickness of the plate to force the metal out (see Figure 29-1). Keep the electrode low, about 5° to 10° from the horizontal plane. Be sure to keep the cut open and prevent molten metal from clinging to the bottom of the plate. Next, soak a few electrodes in water for a few minutes before cutting, then compare the cutting action with the dry electrodes.

Holes may be cut with a 1/8" (3.17 mm) electrode in plates up to about 3/8" (9.52 mm) thick. Use the same setup as in the above procedure. Strike an arc and hold the electrode with a long arc vertically until the metal is molten, then push the electrode through. Once the plate is pierced, move the electrode up and down through the plate and in a circular motion to enlarge the hole (Figure 29-2). If the hole is to be used for a bolt, check the size with an old bolt of similar size. Holes may be cut on heavier metal or cut at a faster rate by using heavier electrodes with a welder of higher capacity. Place metal in a vertical position with the electrode in a horizontal position (Figure 29-3). This will allow the molten metal to run out of the area.

## QUESTIONS

1. What is the amperage range when cutting with electrodes?
2. What forces the molten metal out of the cut area?
3. What type materials can be dangerous to cut?
4. What type of welding electrodes can be used for cutting?
5. What is the approximate angle of the electrode when cutting?

# TIG WELDING-
# TUNGSTEN
# INERT GAS

# chapter 30

Gas tungsten arc welding (GTAW) or Tungsten Inert Gas Welding is often referred to as TIG or Heliarc or Heliwelding. This process uses the heat of an electric arc between a non-consumable tungsten electrode and the base metal.

NOTE: Heliarc is Linde's trade name for TIG as is Airco's Heliweld. The names were derived from the use of helium gas as the shielding gas in the first TIG torches developed (Figure 30-1). GTAW is the AWS identification of this process, but TIG is most commonly used and understood.

A gas, which is inert, usually argon or helium, serves as a shield to protect the molten metal from harmful atmospheric contamination such as oxygen and nitrogen. An inert gas is a gas that does not react with the base or molten metal or the electrode (Figure 30-2).

Additional filler metal, if needed, is fed to the weld area manually. This process is often compared to the oxy-acetylene process where filler rod is added when necessary. In fact, experienced oxy-acetylene welders find TIG rather easy to learn because it requires about the same manual dexterity as oxy–acetylene welding. However, the fundamentals of TIG welding and the welding equipment setup are much more complicated.

TIG welding was introduced about 1943,

and through many improvements of the equipment and the process, it has become a very important method of welding metals that were difficult to weld before—such as aluminum, magnesium, and titanium. Stainless steel, copper alloys, nickel alloys, carbon and low alloy steel, and others can be welded or fused with TIG.

Welding of aluminum by TIG, for example, is easier, and a wider range of thickness can be welded or fused than by metallic arc welding or by the oxy-acetylene welding process. The TIG can produce a high quality weld without the use of flux and therefore eliminates the costly post weld cleanup and possible slag inclusion in the weld area. Because the shielding gas is clear, the welder has a better view of the welding process as compared to the smoke given off by the oxidation of the electrode coating. The welder has more control of the welding process in TIG than in the metallic arc process because the filler rod is fed by hand,

**Figure 30-1** GTAW (TIG) welding.

especially when welding thin gauge metal.

The very high heat of the TIG arc overcomes the high heat conductivity of aluminum, which makes it possible to weld aluminum easier.

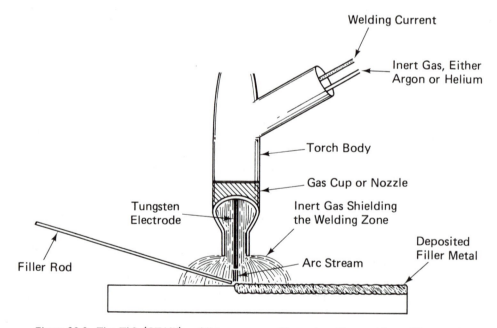

**Figure 30-2** The TIG (GTAW) welding process welds made with or without filler metal. No flux required or flux problems.

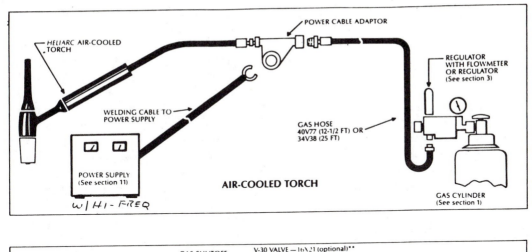

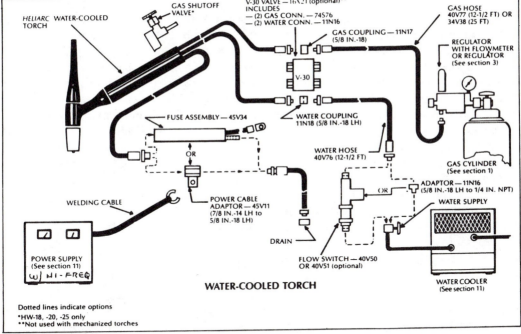

Figure 30-3  Air-cooled torch; water-cooled torch. (*Courtesy of Linde Division of Union Carbide.*)

With TIG, there is very little weld splatter to remove as compared to stick electrodes. The TIG welds are more ductile, corrosion resistant, and stronger than the regular shielded arc or stick electrode welding.

## TIG WELDING EQUIPMENT

The TIG welding equipment with the gas and water controls, high frequency unit, the torch assembly, and remote control (see Figure 30-3) is far more complex than the regular shielded metal arc welding setup. This highly specialized equipment, especially the higher priced units, is easily damaged if not used with care.

When setting up and operating the equipment, the directions or recommendations of the manufacturer should be followed, if they are available. No one should operate the equipment unless they are thoroughly familiar with the equipment and understand how it operates.

## POWER SOURCE

The power source or supply can be supplied by any standard quality AC and/or DC welder with a good current control (Figure 30-4) when used at the lower amperage settings. Some of the AC welding machines and DC machines, which use tap ins or selector dials for current setting selection, may be difficult to use on thin metal because at one setting, the current setting may be too high but the next lower current setting may be too low. The above mentioned machines are the ones where the TIG welding equipment is added or can be added to a regular welding machine not ordinarily designed for TIG welding. Another problem with using these types of welding machines is that the lowest current setting is about 40 amperes, which may be too high for very thin stainless steel.

If a lot of TIG welding is to be done along with stick electrodes, the best choice would be to purchase a machine designed for TIG welding that can be easily switched to stick electrodes (see Figures 30-5a and b).

### HIGH FREQUENCY OSCILLATOR

When using an AC welder for TIG, the AC welder must be equipped with a built-in high frequency unit or be used with an external high frequency unit. There is a problem with the current in the AC welder, which is constantly reversing the direction of the flow of current. When the current changes direction or passes 0, not enough current is flowing, so the arc will become unstable, erratic, or go out. A high-frequency oscillator in the welding equipment, or an external unit, superimposes a high-frequency stabilizing voltage on the welding circuit. This high voltage, over 20,000 volts, ionizes the shielding gas to create an ionized path for the welding current to follow. This ionized path permits making a non-touch start, preferred in TIG welding of aluminum to eliminate the contamination of the electrode and the work piece, if the tungsten electrode should touch the work piece. Non-touch starts are relatively safe in spite of the very high voltage because the current is in milliamperes. The 60-cycle current or 120 changes in direction per second is now up in the hundreds of thousands per second.

Whether the TIG unit is built in or a TIG unit is attached to a welder (Figure 30-6), both units are equipped with a solenoid valve to turn on and off the flow of shielding gas. On some machines, the shielding gas is also used for cooling the torch.

### COOLING THE TORCH

An adjustable after-flow meter or timer is used for cooling the torch and the tungsten electrode after the arc is broken with shielding gas for a brief period, generally in relation to the size of the electrode used (see Figure 30-6).

The more expensive machines or heavy-duty machines are equipped with a water-cooled torch and are likewise equipped with an after-flow meter for cooling the torch and electrode after the arc is broken. The machine using a water-cooled torch can be set up in two different ways. One machine uses a water solenoid activated at the same time as the gas solenoid; the water will flow until the after-

**Figure 30-4** 300 amp constant current power source. (*Courtesy of Miller Electric.*)

flow meter shuts off the shielding gas. Before the welding is started, the solenoid must be activated to check the flow of water.

Another method is to connect the water pump to a 110-volt outlet on the machine, so when the main switch of the welder is turned on, the water pump will start pumping. With some machines mounted on a self-contained water truck, the pump is mounted behind the welder (see Figure 30-5a). The pump generally has a clear bubble glass which will show that water is returning from the torch. If no water is flowing, it must be checked before the welding is started.

## CONTROL DEVICES

Some welders are equipped with remote switches or controls either hand or foot operated. One type of remote control switch, generally hand operated, activates the shielding gas solenoid, water solenoid if equipped, and the high frequency unit after the main switch is turned on. These switches are generally mounted on the torch handle or body. The foot-operated remote control switch also activates the shielding gas solenoid, water solenoid (if equipped), and the high frequency unit. After the main switch is turned on, the

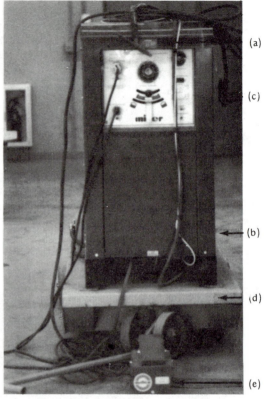

**Figure 30-5A** AC-DC welder with water cooled TIG (GTAW) torch. Output ranges from 5 to 470 amps.

**Figure 30-5B** (a) 500 amp, water cooled torch; (b) circulating pump for the torch; (c) ground clamp; (d) 44 gallon, about 200 liters, tank on wheels for cooling; (e) remote control. Note: Argon cylinder and flow meter-regulator are located behind the welder.

foot pedal is depressed all the way down in order to start the arc. It operates the same way a gas pedal operates on a car—the further down the pedal is pushed, the more power.

When welding aluminum, as the metal heats up, less current is needed for welding; the pedal is partially released lowering the amount of current being used. During some welding operations or when repairing a piece of aluminum or stainless, some parts will require more heat than other parts; with the remote control unit, it is very easy to adjust to the more desirable current. With welders not equipped with remote control units, it is necessary to stop welding and change the current setting on the

**Figure 30-6** Hi-frequency unit that can be used with most AC or AC/DC welders. (*Courtesy of Lincoln Electric.*)

machine. If the welder is some distance from the work area, time is lost when the operator has to make several trips back to the welder to readjust the current setting—especially when several objects are being repaired at the same time.

**High-Frequency Switch** The high-frequency units, either internal or external, have a switch with three positions; one position is off; another position for continuous high frequency used primarily for aluminum; and a start position used for starting the arc, and once the arc is established, the high frequency shuts off. The start position is used for stainless and other metals.

There are many different types of welders designed for TIG and stick electrode; and with the proper equipment they can be used for MIG. Before trying to operate a TIG machine, time should be spent closely examining the various controls so the welder knows what they are there for. Even the best machines can be damaged by improper operation.

# TYPES OF WELDING CURRENT

## *DIRECT CURRENT, STRAIGHT POLARITY*

When welding with direct current straight polarity (DCSP), the tungsten electrode is negative and the base metal or the work piece is positive. As mentioned previously in arc welding, the electrons flow from the electrode to the work piece and two-thirds of the heat of the arc is released in the base metal (Figure 30-7). The weld puddle is deeper and narrower (Figure 30-8) than when DCRP is used. There is less distortion of the work piece because the heat is concentrated in a narrower area and the welds are more rapid because of the heat being released in the work piece. The electrodes can be thinner in diameter because less heat is liberated at the electrode. The positive ion or gas ions flow from the positive side or work piece to the tungsten electrode. These ions, by striking molecules in the atmosphere, create

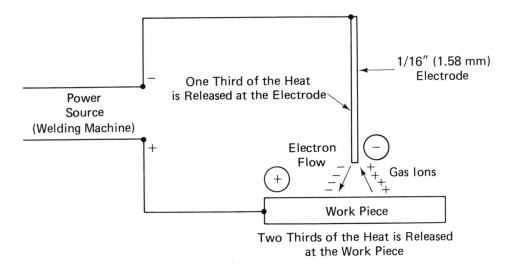

Figure 30-7 DCSP or direct current straight polarity welding.

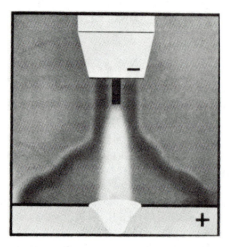

**Figure 30-8** DCSP produces deep weld penetration because the heat is concentrated in the work piece. (*Courtesy Linde Division of Union Carbide.*)

an ionized layer of gas which protects the major flow of electrons or provides a conductor for the flow of current from the electrode to the work piece. Because DCSP does not perform a cleaning action as in DCRP, the work piece must be clean and free of oxides.

## DIRECT CURRENT, REVERSED POLARITY

When using DCRP current, the electrons flow from the work piece to the tungsten electrode. About two-thirds of the heat of the arc is released or generated at the electrode (Figure 30-9). Because of the high heat being absorbed by the electrode, the electrode must be much larger than those used in DCSP. A 1/4" (6.315 mm) electrode would be required where a 1/16" (1.58 mm) would be sufficient for DCSP for the same current setting. The electrons flow to the electrode from the work piece. The positive or gas ions bombard the surface of the work piece breaking up the oxides on the surface. The welds from DCRP are shallower and wider (Figure 30-10) than those from the DCSP. The DCRP should never be used, except on rare occasions, on either aluminum or magnesium. Because of the high heat concentrated at the electrode, the electrodes tend to erode or melt off, and the electrode material may contaminate the weld causing it to crack.

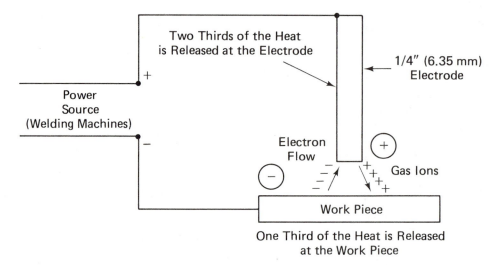

**Figure 30-9** DCRP or Direct current reverse polarity welding.

## ALTERNATING CURRENT

As previously mentioned, the current in AC is constantly changing direction. A sine curve or one complete AC cycle (see Figure 8-4), illustrates that when the current goes above the 0 line, it would be positive, or reverse polarity, and when the current flows below the 0 line, the current is negative or straight polarity because it is going in a different or opposite direction. Therefore, AC current is a combination of DCRP and DCSP. The heat for welding would be produced on the negative or DCSP part of the cycle. The cleaning action would be on the positive part of the cycle. However, if the work piece becomes oxidized, the current would be partially or fully blocked. The current would flow in one direction only, causing the arc to go out, which is called rectification. It is basically the same principle that happens when AC current is rectified for DC welding.

With the use of the high-frequency unit, the positive part of the cycle breaks up the oxides, permitting the current to jump the gap between the work piece and the electrode, and complete

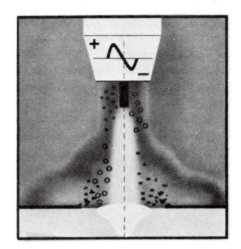

Figure 30-11 The heat is produced on the negative part of the cycle and the cleaning action on the positive part of the cycle. (*Courtesy of Linde Division of Union Carbide.*)

the cycle. The ACHF current produces a deep penetration, wide weld. (Figure 30-11).

NOTE: There is a possibility of radio interference when using high frequency unless the equipment is properly grounded.

## TIG TORCH

The TIG torch is constructed to carry both the welding current and the shielding gas to the work or weld area. Air-cooled torches are designed for light-gauge welding and low amperage settings (Figure 30-12). Welding at high amperage will cause the torch to burn out because the shielding gas is used for cooling and is not adequate at high amperage settings.

The heavy-duty torch is water-cooled; the water circulates around the torch during welding and after the arc is broken to properly cool the torch and the electrode. To conserve gas, some TIG torches are equipped with a shut-off valve to control the flow of gas when not using

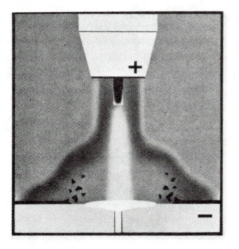

Figure 30-10 Welds from DCRP are shallower and wider than DCSP because most of the heat is concentrated in the electrode. (*Courtesy of Linde Division of Union Carbide.*)

the H/F unit. On other torches, the gas is controlled by the H/F unit. It is important to check the water flow before welding to insure adequate cooling. If the flow has stopped, the torch or welding cable will burn out after a short time. Water-cooled torches require more care during cold weather or in cold climates.

All the cables carrying the welding current, shielding gas and water (if water cooled), are plastic and can be damaged by hot metal or heavy objects falling on them. If necessary, drape the cable assembly over your shoulder to avoid damage. Leather shields or like materials are available to protect the cables.

The tungsten electrodes, called non-consumable electrodes, are held in the torch by a collet and collet body. The collet body is

threaded into the torch body on most TIG torches (Figure 30-13). The collet slips inside the collet body and the tungsten electrode passes through the collet. The gas cap screws in back of the torch body which in turn, tightens the electrode in the collet and collet body of the torch. The collet and collet bodies are made in various sizes to accommodate different sized electrodes. The gas caps are made for different length tungsten electrodes. Check the 0 rings in the gas cap; if worn, replace. A defective 0 ring can cause air to be drawn into the gas flow and

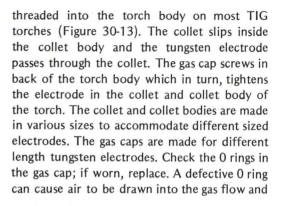

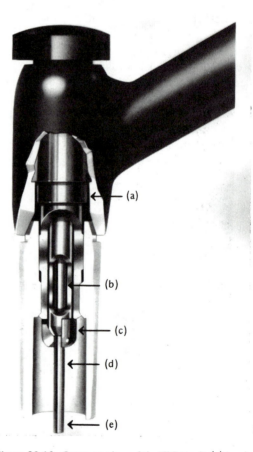

**Figure 30-12** Light duty air cooled TIG torch with long cap for 7" (177.8 mm) electrodes. (*Courtesy of Linde Division of Union Carbide.*)

**Figure 30-13** Cutaway view of the TIG torch; (a) torch body; (b) collet; (c) collet body; (d) gas cup; (e) electrode. (*Courtesy of Linde Division of Union Carbide.*)

cause contamination of the weld and the electrode.

A gas cup or nozzle is threaded on the collet body or into the torch body to give direction to the flow of shielding gas. The cups are available in ceramic and high impact cups. The cups are fragile and care must be taken not to chip or crack them. Any flaws in the opening will affect the proper flow of shielding gas. Some torches are equipped with gas lenses or (Figure 30-14) screens to break up the turbulence of the flowing gas, which may draw in air and cause contamination of the weld. The cups are available in many different diameters and lengths for different sized torches, different electrodes being used, and type of welding operation.

The tungsten electrodes are adjusted by loosening the gas cap and adjusting the electrode to extend out beyond the edge of the gas cup. The length of the extension beyond the edge of the cup depends on the type of welding to be done and the type of joint. For welding on flat positions, the tungsten electrode should stick out about the width of the electrode (Figure 30-15) or about 1/8" (3.17 mm). For welding inside corners or fillets, the electrode will extend further out of the cup, or up to 1/4" (6.34 mm) (Figure 30-16).

## SHIELDING GAS

The shielding gases are argon and helium or a combination of both and are called inert gas because they are chemically inactive. Argon is the most commonly used inert gas because it is cheaper and it provides a more effective shield. Argon is about ten times heavier than helium (Figure 30-17) and one and a half times heavier than air. Argon permits a smoother and better controlled arc, lower voltage for thin metal, and better cleaning action when welding aluminum and magnesium. Helium is more frequently used for heavier metal, for high speed welding, and machine welding. Argon is often mixed

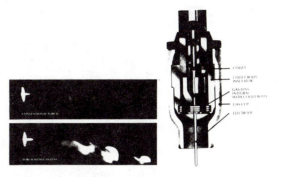

**Figure 30-14** Gas Lens. (*Courtesy of Linde Division of Union Carbide.*)

with helium for machine welding. Because helium is seven times lighter than air, helium floats away from the weld area very quickly and is affected by any air currents in the welding area.

The two gases, argon and helium, are supplied in steel cylinders varying from about 80 to 330 cubic feet (2320 to 9570 liters) at a pressure of about 2200 lbs. per square inch (15,180 kPa) at 70°F (21.1°C), or room temperature. A regulator must be used to reduce the high cylinder pressure to a usable working pressure. A combination regulator and flowmeter is used in TIG welding. These regulators are similar to those used for oxy-acetylene welding but the gas is not measured in pounds per square inch or kPa. Rather it is measured in cubic feet per hour (CPH) or liters per minute (LPM). The device for measuring the flow of gas is called a flowmeter (Figure 30-18) which provides better control of the gas flow. On some regulators the pressure and the flow are adjusted by a valve on the flowmeter. Some control the flow by turning in the adjusting screw on the regulator. The height of the ball in the glass tube indicates the amount of flow. The amount of flow will vary with the thickness of the metal to be welded. The regulator flowmeter combination is available in either single- or two-stage regulators. The flowmeter must be kept in a vertical position to indicate accurately the amount of

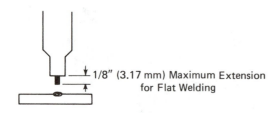

**Figure 30-15** Electrode extension.

gas flow. The flowmeters are delicate instruments to be handled with care.

## ELECTRODES

The non-consumable electrodes are available in several different compositions, in diameters from .020″ to 1/4″ (.50 mm to 6.350 mm) and lengths from 3″ to 12″ (76.2 mm to 304.8 mm). The pure tungsten electrodes (a purity of 99.5%) have a high current capacity with very little electrode consumption; they are generally coded green. The pure tungsten is often preferred for AC welding if aluminum is being welded because it provides a desired balling characteristic for arc stability and weld contamination is eliminated.

The thoriated tungsten electrodes, with a range of 1% to 2% thorium added, permit easier arc starting and increased current carrying capacity by about 50% over pure tungsten.

NOTE: 1% thorium electrodes are generally coded yellow.

This also permits a wider operating current

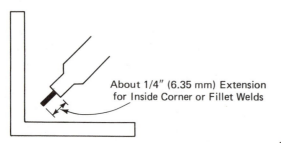

**Figure 30-16** Electrode extension.

| GAS | ARGON | HELIUM |
|---|---|---|
| Density* | 0.363 lb/ft³ (5.815 kg/m³) | 1.04 lb/ft³ (16.659 kg/m³) |
| Flammable Limits in Oxygen | – | – |
| Flammable Limits in Air | – | – |
| **LIQUID** | | |
| Density* | 86.98 lb/ft³ (1.39 g/cm³) | 7.798 lb/ft³ (0.125 g/cm³) |
| Heat of Vaporization | 70.2 Btu/lb | 8.72 Btu/lb |
| Boiling Point† | −302.6 F (−185.9 C) | −452.1 F (−268.9 C) |
| Molecular Weight | 39.95 | 4.0026 |

*At boiling point
†At 1 atmosphere

**Figure 30-17** Comparison chart of argon and helium. (*Courtesy of Linde Division of Union Carbide.*)

range for a specific diameter of electrode. It is well suited for DCSP welding current, with a higher arc stability. The 2% thoriated electrode is used for thin sheet metal welding such as

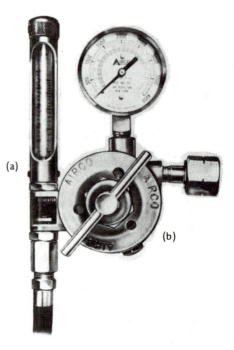

(a)

(b)

**Figure 30-18** Argon and helium regulator-flowmeter; (a) flowmeter; (b) regulator. (*Courtesy of Airco.*)

stainless and will maintain the formed tip longer than the 1% thoriated electrode generally coded red. Under certain conditions, thoriated electrodes can be used for AC welding of aluminum. The zirconium tungsten electrodes are similar to thoriated but they have increased electron emission and current carrying capacity provided by the addition of zirconium. A striped tungsten electrode has a strip of thorium imbedded the length of the electrode. This combination gives the desired balled effect and the ease of arc starting. The diameter of the electrodes (Figure 30-19) used is determined by the metal thickness and the amount of current to be used.

The electrodes used for welding such metal as stainless with DCSP current must be ground to an even point (Figure 30-20). When welding DCSP, a worn-off point could cause the arc to wander or become erratic. Stop and regrind to a straight, even point. When using AC/HF with aluminum, the electrode should have a balled or rounded end (Figure 30-20). The ball will form by itself with welding. If the ball becomes too large, this could be an indication of excessive current. Keep the electrode straight and in the center of the cup opening. Otherwise the arc

will not be in the center of the gas shield. The electrodes are easily bent when red hot and this could happen when laying the torch on a bench. If the electrode becomes contaminated or coated near the end, remove or break the tip.

NOTE: Always keep the color-coded end inside the torch to avoid mixing up the different electrodes.

## FILLER METAL RODS

Filler rods used for TIG welding should match the type of metal to be welded. But on many occasions, the exact composition of the metal to be welded is unknown such as when repairing objects made of aluminum, stainless steel, or other metals. When sheet material of aluminum, stainless steel, or other metals, is to be welded, the sheets are generally marked with the type of alloy contained in the sheets. Therefore, it is easier to determine what type filler rod to use.

The aluminum filler rod, ER4043, generally produces good results when repairing aluminum objects such as transmission cases, lamp post brackets, extruded window frames, and the

### Recommended Electrodes, Cups and Metal Nozzles for Various Welding Currents

| Electrode Diameter | | Use Cup or Metal Nozzle No. | Welding Currents, Amp | | | |
|---|---|---|---|---|---|---|
| | | | ACHF* | | DCSP** | DCRP** |
| in. | mm | | Using pure tungsten electrodes | Using thoriated electrodes† | Using pure, or thoriated tungsten electrodes | |
| .020 | 0.5 | 4, 5 | 5-15 | -- | -- | -- |
| .040 | 1.0 | 6 | 10-60 | 60-80 | 15-80 | -- |
| 1/16 | 1.6 | 6 | 50-100 | 100-150 | 70-150 | 10-20 |
| 3/32 | 2.4 | 6, 8 | 100-160 | 160-235 | 150-250 | 15-30 |
| 1/8 | 3.2 | 8 | 150-210 | 225-325 | 250-400 | 25-40 |
| 5/32 | 4.0 | 8 | 200-275 | 300-425 | 400-500 | 40-55 |
| 3/16 | 4.8 | 8, 10 | 250-350 | 400-525 | 500-800 | 55-80 |
| 1/4 | 6.4 | 10, 12 | 325-475 | 500-700 | 800-1100 | 80-125 |

All current values are metered readings. Most transformers deliver about 15 percent more current than shown on their scale readings.

The maximum current values shown in the table for ACHF have been determined using an unbalanced wave transformer. If a balanced wave transformer is used, either reduce the maximum values in the table by about 30 percent or use the next larger size electrode. This is necessary because of the higher heat input to the electrode in a balanced wave setup.

Thoriated tungsten electrodes are recommended when a gas lens and high frequency starting are used.

Balled electrode tip ends can best be formed and maintained at these ac current levels.

**Figure 30-19** Recommended electrodes, cups and metal nozzles for various welding currents. (*Courtesy of Linde Division of Union Carbide.*)

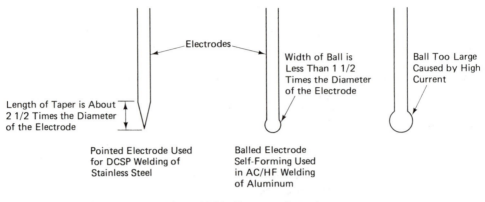

Figure 30-20   Tungsten electrode.

like. Many objects made of stainless steel can be repaired by using ER308ELC stainless steel filler rods. This rod can also be used for oxy-acetylene welding of stainless steel.

Mild steel, copper-coated, welding rods as used for oxy-acetylene welding cannot be used for TIG welding because of contamination from the copper coating. Bare steel rods containing deoxidizers can be used for TIG welding, similar to AWS E705-2.

## ALUMINUM WELDING WITH TIG

Extreme care must be exercised when welding with a TIG welder because of the complicated and costly mechanism involved. The lack of sufficient knowledge and/or carelessness in setting up the welder for welding can be costly. The recommended procedures must be followed very closely in order to produce satisfactory results. Using the wrong type of current or excessively high currents, failure to turn on the shielding gas or checking the water flow, can ruin the torch assembly very quickly.

Use 1/8" to 3/16" X 3" X 6" (3.175 mm to 4.760 mm X 76.2 mm X 152.4 mm) aluminum plates, an AC welder with high frequency with argon or complete TIG welding outfit, 1/8" (3.175 mm) 4043 welding rod, and a stainless steel brush. This procedure can be done with either 1/16" (1.58 mm) or 1/8" (3.175 mm) electrodes. Set up the welding machine. Check the condition of the electrode and if it is contaminated, remove the end and adjust the electrode to the proper stickout from 1/8 inch, or equal to the diameter of the electrode as recommended by some manufacturers. Press the electrode against a solid surface for tightness. If the electrode moves, tighten the gas cap or collet holder. Check the manufacturer's recommendations or guide chart, if available. Set the machine to the proper welding amperage. Turn on the shielding gas and set to the proper flow, generally 12 to 15 CFH (cubic feet per hour) (6 to 7 LPM) for 1/16" (1.58 mm) and 15 to 20 CFH (7 to 10 LPM) for 1/8" (3.175 mm) electrode. If the torch is water cooled, turn the water on and check water flow.

Clean the pieces of aluminum to be welded by brushing with a stainless steel brush, scraping, or filing. Avoid grinding the aluminum because some abrasive material may be imbedded in the weld area. Position the pieces for welding on the work table. Attach the ground cable securely and turn on the welder. Turn on the HF unit and recheck the gas flow and the water cooler, if it is water cooled. To start the arc, turn the torch in horizontal position with the work piece, with the cup or nozzle barely touching the work piece (Figure 30-21). Press the remote or start button. The HF current will jump the gap between the electrode and the work, establishing the arc. Immediately, raise the torch to a $75°$ to $85°$ angle from the horizontal or $15°$ to $25°$ from the vertical (Figure 30-22) with the electrode about 1/8" (3.17 mm) from the surface. Holding the torch at an excessive angle can lead to contamination of the air being siphoned into the argon gas shield. If the arc does not start without touching, the HF spark intensity may be increased.

## ALTERNATE METHOD OF STARTING THE ARC

With the welder turned on, hold the torch vertically about 1" (25.5 mm) above the surface of the work piece. Start to lower the torch and press the remote or start button. When the

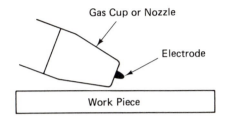

**Figure 30-21** Starting the torch. Hold torch in this position with gas cup or nozzle barely touching the work piece. Press remote or start button.

torch is about 1/8" to 1/4" (3.17 mm to 6.34 mm) (Figure 30-23) from the work piece, the HF current should jump the gap to establish the arc. Another method is to hold the torch horizontally or parallel about 2" (50.8 mm) above the work piece (Figure 30-24). With the welder turned on, press the remote or start button and swing the torch to a vertical position with the electrode within 1/8" (3.17 mm) of the work piece. The HF current should jump the gap and establish the arc.

Perhaps it would be better to make a few practice swings before pressing the remote or

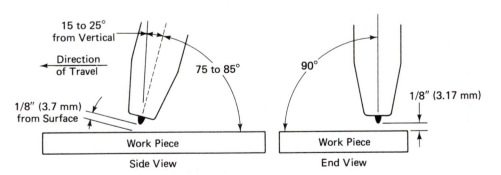

**Figure 30-22** Hold torch $75°$ to $85°$ in the direction of travel.

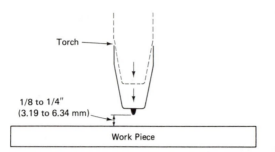

**Figure 30-23** Hold torch 1" (25.5 mm) above surface. Slowly lower the torch to 1/8" to 1/4" (3.17 to 6.34 mm) above work piece and press start button. The arc will jump the gap.

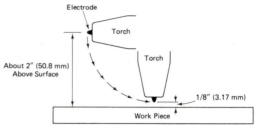

**Figure 30-24** Hold torch in horizontal position. Press start button and swing torch to a near vertical position to start the arc. Alternate methods of starting the torch.

start button to get the general idea. At no time should the electrode be permitted to touch the work piece when welding aluminum. When using DCSP on stainless steel or other metals, the electrode must touch the work in order to start the arc. This will be discussed in the next section.

Hold the torch in one position until a wet spot or puddle forms under the electrode. Add filler rod by dipping the rod in and out of the puddle, and move the torch forward (Figure 30-25), tilting the torch about 15° to 25° in forward direction. Keep the filler rod at a low angle, about 10° to 15° from the work piece. NEVER allow the filler rod to touch the electrode. If this happens, stop, remove the contaminated end, and readjust the electrode with the welding machine turned off. Run the bead across the work piece, adding enough filler rod to form a slightly raised bead. Too much filler rod will result in a bead that is too high. Run another bead across the plate. If the plate becomes too hot, the current may be reduced. If not, excessive heat will cause the plate to become soft, especially when welding near the end of the plate. If the welder is equipped with a foot control, the heat can be controlled by less pressure on the foot control. Observe the contour and width of the bead. This type of welding rod and torch manipulation is similar to the oxy-acetylene welding technique.

In many cases, the amperage can be set lower as recommended by the manufacturer or by the guide chart by taking into consideration the total surface area and the thickness of the aluminum to be welded. At lower amperage, the initial formation of the puddle or wet spot takes a little longer, but the work piece will not get as hot as the weld proceeds.

## WELDING JOINTS

### WELDING BUTT JOINTS

Use two pieces, 1/8" to 3/16" X 2 X 6" (3.175 mm to 4.763 mm X 50.8 mm X 152.4 mm) aluminum plates, AC/HF welding outfit complete, a stainless steel brush or file, and 1/8" (3.175 mm) 4043 aluminum welding rods.

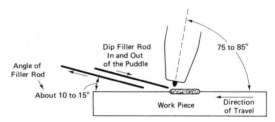

**Figure 30-25** Method of adding filler rod.

Clean the aluminum plate to be welded by brushing or filing. Position about a 1/16" (1.58 mm) gap between the two pieces. Check the electrode for condition. If it is contaminated, grind it off and readjust it to proper stickout. Follow the same procedure as was previously stated for operating the TIG welder. Check the water flow, if water cooled. Tack weld the pieces together, start to run a bead at one end and run the length of the joint holding the torch at a 15° to 25° angle in the direction of travel. Be sure to get adequate penetration or melting of both edges of the joint. Add the necessary filler rod and avoid too much build-up (see Figure 30-25). Observe the bead and check the penetration on the root of the joint.

## LAP WELDING OF ALUMINUM

When lap welding, applying too much heat to the top plate will cause the edge of the top plate to melt away before the bottom plate puddle is formed. Holding the torch at an approximate angle of 65° to 80° from the horizontal or 15° to 35° from the vertical (Figure 30-26) will direct more heat to the bottom plate. During lap welding, the puddle on the two plates will form a V effect, sometimes called a notch. The V or notch (Figure 30-27) will travel with the torch if the torch is not moved too fast. Be sure that the V or notch

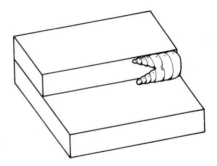

**Figure 30-27** V or notch formed in the molten puddle.

is completely filled the entire length of the bead.

Add only enough filler rod or material to obtain proper bead formation. The amount of filler material to add each time the rod is dipped in and out of the puddle depends on the diameter of the rod being used. It is very easy to add excessive filler material. Holding the rod in the puddle too long will cause excessive amounts of filler rod to be melted. Avoid melting the filler rod on a cold or unmelted surface. This happens when the rod is held too high or remains too close to the surface. When preparing metal for lap welding, it is very important to clean the electrode, clean off the edge of the top plate, and the top surface. Oxides left on the edge could hinder the puddle formation and could introduce impurities into the bead.

Use two pieces, 1/8" to 3/16" X 2" X 6" (3.178 mm to 4.762 mm X 50.8 mm X 152.4 mm) aluminum plates, AC/HF welding outfit complete, stainless steel brush or file, and 1/8" (3.175 mm) or 5/32" (3.969 mm) 4043 aluminum filler rod. Clean the aluminum plates to be welded by brushing or filing. Position the plates. Check the electrode for condition. If contaminated, grind it off and readjust it to proper stickout. Follow the same procedure as previously stated for operating the TIG welder. Check for water flow, if water cooled. Tack

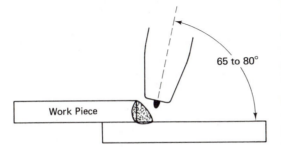

**Figure 30-26** Torch angle for lap welds.

65 to 80°

Work Piece

weld the plates on both ends, and start to run the bead across the lap joint, holding the torch at the proper angle. Add just enough filler rod to form the proper bead, and observe the bead contour and penetration. Turn over the plate and run a lap bead without filler rod. If the plates are too hot, cool them first before continuing or else the plate will become softened or mushy unless the welder is equipped with a foot control. The plates can be cooled with water and wire brush before wleding. For additional practice, run an additional bead over the previous weld, but this time add filler rod.

## FILLET OR INSIDE CORNER WELDS

Fillet welds with TIG are a little different because the gas cup or nozzle restricts the torch from getting in close to the root of the joint. The electrode stickout is increased to about 1/4" (6.350 mm). As in arc welding, avoid undercutting along the edge of the bead. Undercutting is caused by improper manipulation of the torch. Some inexperienced TIG operators may experience difficulty in getting the proper fusion of the two plates. One technique is to form a small puddle on the bottom plate and add a small amount of filler rod, then form a puddle on the vertical plate and add a small amount of rod. By the use of a weaving motion, the two small beads or spots will join together.

Once this is accomplished, continue the bead by weaving back and forth and pausing at the bottom and vertical plate and adding a small amount of filler rod (Figure 30-28). As in lap welding, the bead will form a notch; make sure the notch is completely filled.

Use several pieces of 1/8" to 3/16" X 4" X 6" (3.175 mm to 4.762 mm X 101.6 mm X 152.6 mm) aluminum plates, AC/HF welding outfit complete, stainless steel brush or file, and 1/8" (3.175 mm) or 5/32" (3.969 mm) 4043 aluminum filler rod. Clean the aluminum plates. Check the electrode for condition and adjust it to about 1/4" (6.35 mm) stickout (Figure 30-29). Follow the same procedure as previously described for operating the TIG welder. Check the water flow, if water cooled. Tack weld a 4" X 6" (101.6 mm X 152.6 mm) plate vertically on a 4" X 6" (101.6 mm X 152.2 mm) plate at the ends. Start the bead and continue across the joint, holding the torch at about 40° (Figure 30-30) from the vertical to the joint and about 20° forward from the vertical. Add just enough material to produce a satisfactory bead. Hold the filler rod at a low angle to avoid contact with the electrode and to avoid melting too much filler rod at one time. Observe the contour of the bead. When the plates are cool enough, run a bead on the opposite side of the joint. If the welder is foot or remote controlled,

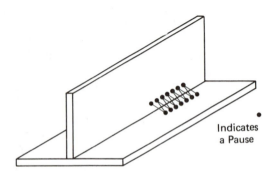

Figure 30-28 Weaving motion with a pause on the two plates.

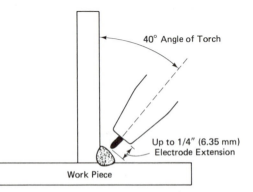

Figure 30-29 Inside corner or fillet welds.

cut back on the amperage before starting on the opposite side.

NOTE: The plates can be cooled with water, followed by wire brushing, before welding.

## DIFFICULT WELDING POSITIONS

### *HORIZONTAL WELDING WITH TIG*

Welding in a horizontal position with TIG is very similar to welding in a flat position, but with the force of gravity working on the molten metal. By keeping the puddle from getting too large and by adding only enough filler rod at one time to form a satisfactory bead, horizontal welding is quite easy. Holding the torch about 15° from the horizontal will help to prevent undercutting of the upper edge of the bead. The filler rod should be added to the puddle from the upper side. The TIG operator has more control of the puddle (as in torch welding) than the stick electrode welder.

### *VERTICAL WELDING WITH TIG*

TIG welding in a vertical position requires more practice but is not too difficult to master. On thin aluminum sheets, the vertical down method produces adequate welds without burn through. Be careful to avoid large puddles and adding excessive filler material to prevent the molten metal from sagging under the force of gravity. The torch is normally held at about a 90° angle to the weld surface with the filler material being added to the forward edge of the puddle. Keep the filler rod at a low angle. Vertical up welding provides more penetration and is used for welding heavier sheet and plate aluminum. The size of the puddle must be controlled, and only the necessary amount of filler material should be added to the puddle to prevent the molten metal from running down. The torch should be held upward at an angle of about 60° from the horizontal (Figure

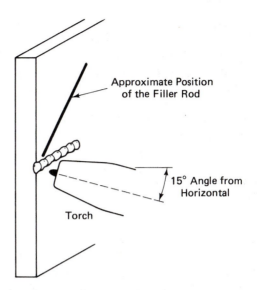

**Figure 30-30  Horizontal welding.**

30-31). The filler material is added to the forward edge of the puddle or the top.

### *OVERHEAD WELDING*

Overhead welding requires more control of the puddle and the amount of filler material to be melted in the puddle because of the force

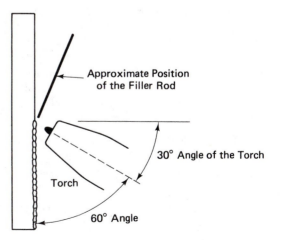

**Figure 30-31  Vertical welding.**

of gravity. A smaller puddle and consequently a smaller bead is recommended for overhead welding. Further control can be accomplished by decreasing the amperage to that used in flat welding. Amperage depends on thickness of metal and size electrode being used. The torch and filler rod are held at about the same angles to that of flat welding.

## WELDING MAGNESIUM WITH TIG

Welding magnesium with TIG is similar to welding aluminum. Magnesium is silvery white, lighter than aluminum in weight (approximately two-thirds as heavy as aluminum). Magnesium shares several characteristics with aluminum such as high heat conductivity, rapid oxidation, low melting point (about 15°F lower than aluminum), and a high thermal expansion. The same equipment, techniques, and joint preparations are used for both metals.

Magnesium is not strong in the pure state but when alloyed with other elements, it is quite strong. Magnesium alloys are used in aircraft manufacture, some automobile casings, or where weight is a factor. Example—steel is about four and a half times heavier than magnesium.

As mentioned in the aluminum section, DCRP is rarely used except to weld magnesium occasionally. When using DCRP, the electrodes must be larger or at least twice as large as compared to ACHF. DCRP can be used with either argon or helium. The resulting welds are wider and shallower.

NOTE: If helium is used, the gas flow is twice as much as argon. DCRP current provides a good cleaning action of the work piece similar to ACHF.

A shorter arc length is recommended for magnesium than used for aluminum.

## WELDING STAINLESS STEEL WITH TIG

One of the most commonly used alloys is stainless steel. Stainless steel offers a combination of mechanical properties, corrosion resistace, and heat resistance unmatched by any other commercial metal. It is found in many household appliances and utensils such as pots, pans, toasters, or stoves, food processors and in chemical manufacturing equipment or any other objects or structures that require those properties provided by stainless steel.

### COMPOSITION OF STAINLESS STEEL

Stainless steel contains at least 11% chromium, which makes the stainless steel stainless or corrosion-resistant. Most stainless steels contain from 16% to 24% chromium. The higher the chromium, the greater the corrosion resistance and resistance to oxidation at high temperatures.

Another metal, nickel, in excess of 6%, increases the corrosion resistance and improves the mechanical properties of stainless steel. Other elements are added to stainless steel in smaller quantities to give stainless additional desirable properties. For example, molybdenum is added to resist pitting-type corrosion; sulfur and selenium impart free-machinery characteristics.

When repairing objects made of stainless, it is practically impossible to tell the exact composition of the stainless steel. Some of this information is available from the manufacturers, but usually there is not enough time to secure the information needed from the manufacturer.

The filler used for welding should be about the same composition of the object to be welded. There are about 44 different grades or compositions of stainless steel, but some filler rods can be used for a number of grades like AWS ER 308ELC or ER 347. Stainless steel is directly the opposite of aluminum in thermal or heat conductivity. The heat generated during welding does not spread out as fast as in aluminum or steel. The spread of heat from the weld area is about 50% less than steel and several times less than aluminum. While the heat conductivity is less, the heat expansion is twice as much as steel. These two factors, the lack of heat conductivity and heat expansion, are the causes of warping of the base metal when welding thin gauge materials. This is why stainless steel thin gauge material must be clamped down or placed in holding devices to prevent warping. Clamping devices made of copper or copper backup bars are excellent for controlling heat because copper has a high rate of thermal conductivity. Welding is not difficult and requires less heat due to thermal conductivity because the heat stays in the weld area longer.

## WELDING CURRENT AND ELECTRODE

Stainless steel can be welded with either direct current, straight polarity (DCSP) or alternating current, high frequency (ACHF). DCSP is recommended on heavy gauge material because of higher heat input resulting in deep penetration and higher rate of speed. ACHF is recommended for thin gauge material because of lower heat input. When using ACHF, the high frequency is used for starting only or non-touch start. When welding with DCSP, the electrode must contact the surface to establish the arc, which could possibly contaminate the electrode. The recommended electrode for stainless steel is 2% thoriated tungsten electrode. Due to the lower amperage required for welding stainless steel, a small diameter electrode can be used. For example, material up to about 3/16" (4.76 mm) can be welded with a

1/16" (1.58 mm) electrode.

The electrode must be sharpened to a point with a taper about two and a half times the diameter of the electrode (Figure 30-20). When sharpening the electrode, the point must be dead center or else the arc will be off center and will not be in the center of the gas shield. When the point becomes worn, the arc may wander or become erratic.

Most electrodes are color-coded, therefore the point should be made on the opposite end so the electrode can be identified—especially for thoriated and pure tungsten electrodes. Generally, pure tungsten is coded green, 1% thoriated tungsten is coded yellow, and 2% thoriated tungsten is coded red. The amount of electrode stickout is the same as aluminum, remembering the different distances for flat and fillet welds.

## WELDING PROCEDURES

To develop skills in stainless steel welding, use thinner gauge metal rather than heavier metal. The reason for this is that thinner gauge metal, like 20 gauge (1.2 mm), will warp easier, the danger of burn through or melting away the edges is greater, and the welds are naturally smaller, so problems will show up faster. With heavier gauge metal, such as 10 or 12 gauge (about 3.5 mm or 2.6 mm) and heavier, it is easier to control the weld material; with reasonable judgement there is less burn through or melting away the edges; and the heavy metal does not warp as easily. If excessive heat and weld material are used without backup plates and clamps, the metal will still warp.

When repairing objects made of stainless steel like large cooking containers, there is less chance of warpage due to the shape of the object. After a person has gained some experience in welding stainless, objects made of stainless steel that require repairing should be obtained. Repairing is a lot different than welding a clean flat plate or sheet.

**Running Beads With Stainless Steel** Obtain stainless steel sheets about 14 gauge X 6" X 6" (1.75 mm X 152 mm X 152 mm) and remove the oxides by brushing. Prepare an electrode for stainless steel welding, preferably using 2% thoriated tungsten. The point taper should be about two to two and a half times the diameter. Adjust the electrode stickout from 1/8" to 3/16" (3.17 mm to 4.76 mm) depending on the diameter of the electrode. With stainless, the electrode may be flush with the gas cup for flat position. Open the argon cylinder valve. Switch the current selector to Direct Current negative (DCSP), and reduce amperage to 80 to 100 amperes. With some of the larger or higher capacity welders, 60 amperes is sufficient for 14 gauge metal. Turn the high frequency switch to the start only position. Press the remote or start switch and set the argon between 10 to 25 CFH (5 to 12 LPM) depending on the welder. Some manufacturers recommend a higher flow of argon than others. Check the water flow if the torch is water cooled.

Select a 1/16" (1.58 mm) filler rod that matches the composition of the work piece, if known. If not, an ER 347 can be used on many types of stainless or try a ER 308 ELC filler rod. Use the smallest diameter filler rod on thin material. If the rod is too large, it may cool down the molten spot adding excessive filler metal, which could make the bead too large.

Starting from one side of the sheet, strike the arc, using one of the several methods previously described. Hold the torch at about an 80° angle and tip the torch about 10° in the direction of travel with the electrode 1/8" (3.17 mm) from the work piece. Form a small puddle or molten spot, and add the filler material to the edge of the puddle. Move the torch forward, adding filler material by dipping

the end of the rod in and out of the puddle. Add only enough filler material as needed to form the bead. Keep the filler rod at a low angle or about 5° to 15° from the horizontal to avoid touching the electrode. When using 1/16" (1.58 mm) filler rod on thin sheets, keep the bead around 1/8" (3.17 mm) in width. Avoid getting the puddle too large because it may melt the sheet.

If the work piece melts too quickly, reduce the welding current, if manually controlled, or if a remote foot control is used, let up on the pedal. On the other hand, if it takes too long to form a suitable molten puddle, increase the current. Keep the current as low as possible, yet high enough for a satisfactory weld. This reduces the amount of heat in the work piece, thereby cutting down the amount of distortion. Weld across the sheet and cool it. Wirebrush the bead and observe the contour and ripples. Check the underside for fall through. Run several beads across the width of the sheet and observe the results.

## BUTT WELDING

Whenever possible, the metal should be placed in a clamping device with a copper back-up strip directly under the joint to control the heat and prevent distortion. Because only the edges of the sheet are in the weld zone, the edges will become wavy if not clamped. When the edges of the sheets are a close match, the joint can be fused together. When fusing the edges together, a small hole will appear, will travel with the torch, and will be filled in as the torch moves ahead. If the hole gets too large, filler rod must be added. This is sometimes called a keyhole. The problems with fusion welding a butt joint are the possibility of warping due to expansion and the melting away of the edges. Leaving a gap between the sheets will eliminate the warping due to expansion but filler rod must be used.

When welding butt joints on heavier materials, the edges must be bevelled similar to steel plates for adequate penetration and fusion. Use the smallest diameter filler rod to avoid cooling the molten puddle; avoid excess filler material in the bead which makes it more difficult to form the correct bead.

## LAP WELDS

Like any other lap joint weld, the top piece has a tendency to burn away more quickly. More heat should be directed to the bottom piece. Like aluminum, the lead edge of the bead will form a notch or V shape effect. Add the filler rod into the puddle nearest to the top plate to fill in the V and fill in the lap piece if the edge starts to melt back. Keep adding filler metal at an even rate, dipping the rod in and out of the molten puddle. Lap joints will require more filler metal than a flat weld. It may be necessary to clamp the two pieces together, if the two pieces are tight together and cannot warp or separate. Careful movement of the torch is required so as not to melt back a section of the upper piece.

## FILLET WELDS

With a fillet or inside corner, the same problems exist where the vertical plate edges have a tendency to melt away before the flat piece. The torch should be concentrated more on the bottom plate. When the puddle forms on the bottom plate, filler should be added to the puddle nearest to the vertical piece. This will also cut down the amount of undercutting of the vertical piece. The electrode must be extended out farther for fillet welds because the gas cup or nozzle restricts the torch from getting in close. The electrode should stick out about 1/4" (6.35 mm) but could be more depending on the diameter of the cup or nozzle. The metal should be clamped in position and copper backup strips should be used, whenever possible.

If strength is required, it may have to be welded on both sides. Welding in different positions is about the same with most welding processes, but stainless steel welding has the advantage of lower heat being used—the molten puddles are smaller and more controllable.

## QUESTIONS

1. What are some of the trade names for gas tungsten arc welding?

2. What is the GTAW or TIG process?

3. In TIG welding, what are the two inert gases generally used as shielding gases?

4. What is meant by inert gas?

5. What are the three metals that are easier to weld with TIG than with the metallic arc?

6. When welding aluminum, what are two advantages of TIG welding versus oxy-acetylene and metallic arc welding?

7. What is one feature of the TIG arc which makes it easier to weld aluminum?

8. What type of power unit should be used for TIG welding?

9. What is the disadvantage of using tapin or selection dial type power units?

10. When using an AC machine, what type unit must be added?

11. What are two important purposes of this unit?

12. What is the purpose of the after-flow meter or timer?

13. What type additional cooling is used on heavy duty machines and torches?

14. What is the advantage of foot-operated remote control units?

15. What is the three-position high frequency switch used for?

16. Why is it necessary to use heavier tungsten electrodes for DCRP welding?

17. What holds the electrode in the torch body?

18. What is the purpose of the gas cup or nozzle?

19. What is meant by electrode stickout?

20. What are some of the differences between argon and helium gas?

21. What is used to regulate the argon or helium gas?

22. What are some of the different types of electrodes?

23. What type of electrode point is used for welding stainless steel and aluminum?

24. What type of filler metal or rod should be used for aluminum or stainless steel?

25. How should aluminum be prepared before welding?

26. On which metal, stainless steel or aluminum, can the touch start be used?

27. At what angle should the filler rod be held?

28. What is the average stickout for welding butt and lap joints?

29. What length stickout is necessary for inside corner welds?

30. When horizontal welding, where should the filler rod be added?

31. Which type vertical welding is recommended on the aluminum plates?

32. What is the chromium content of most stainless steels?

33. What is another metal that is added to make stainless steel more corrosion resistant?

34. What is the thermal conductivity of stainless steel as compared to aluminum?

35. What causes stainless steel to warp easily?

36. What can be used to prevent warping while welding thin gauge stainless steel?

37. What type electrode is recommended for stainless steel?

38. What happens to the edges of the thin plates when butt welding with stainless?

39. When lap welding, where should most of the heat be directed to avoid melting away the edge of the top plate?

40. Why is it important to use a smaller diameter filler rod?

# GAS METAL ARC WELDING OR MIG

# chapter 31

The gas metal arc welding (GMAW) process more commonly called MIG (metallic inert gas) uses a consumable electrode in the form of a continuous wire. GMAW is the AWS identification for the process but MIG has been the long-standing reference. Like the TIG process, the molten metal in the weld area is shielded by an inert gas or gas mixture. The MIG welding process offers many advantages over electric arc welding. Some of them are: a high speed welding and metal deposit; less distortion and a narrow heat or weld zone; no slag to remove; little waste as material has no stubs to throw away; continuous welding process without changing electrodes; and high quality welds at low cost.

The system is semi-automatic; the current setting, gas flow, and wire feed are pre-set. The torch is operated manually. For production purposes, the MIG process is fully automatic. The MIG process can be used for welding various types of steel, aluminum, stainless steel, and other metals in all positions.

Metal from about 1/32" (.76 mm) (Figure 31-1) to heavy plate (Figure 31-2) can be welded with MIG. The automotive industry has used MIG in the production line for assembling bodies but is now increasingly using MIG

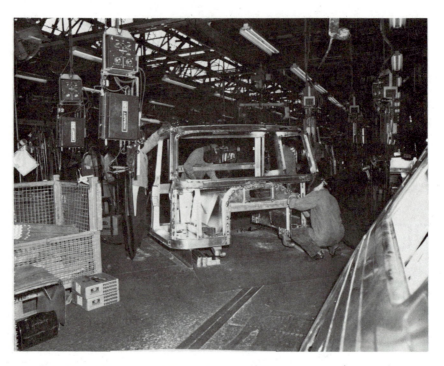

**Figure 31-1**  Using MIG for assembling a cab made of light gauge metal. (*Courtesy of Hobart Brothers.*)

welding in repairing of the new type bodies.

NOTE: Automobile manufacturers use a HSS (high strength steel) in the underbody parts in some of their smaller automobiles and recommend MIG for repairing and replacing the underbody parts. The second choice would be stick electrodes, but one should avoid oxy-acetylene.

Different manufacturers use trade names for the MIG welding process and equipment such as Airco's Aircomatic and Dip Transfer, Hobart's Micro wire and Systematic Fender mender, plus many more.

## TYPES OF MIG WELDING

There are several different types of MIG, with the spray arc and the short arc, or fine wire, being the most common outside of those different types used in industrial plants or production situations.

## *SHORT ARC PROCESS*

In the short arc or short circuiting metal transfer (Figure 31-3), the electrode metal transfer occurs when an electrical short is established. The short arc process occurs in several stages or steps. In the initial stage, the electrode wire contacts the work piece (Figure 31-4), creating an electrical short circuit for a fraction of a second. The heat generated melts a small piece of the wire into a globule. The electrical current pinches the globule, which falls to the work piece. The arc is ignited, and the base metal, or work piece, is melted to form a pool along with the globule. At the same time that the electrode wire is advancing to create another short circuit, the arc will heat up the end of the wire and start to melt it. The arc

Figure 31-2 Heavier metal being welded with MIG (GMAW). (*Courtesy of Miller Electric*.)

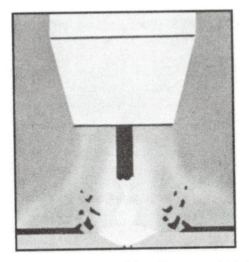

Figure 31-3 Short arc welding. (*Courtesy of Linde Division of Union Carbide*.)

will be extinguished momentarily until the globule is pinched off, and the arc will be reignited.

The metal from the electrode is only transferred to the work piece during the short-circuiting period and not during the arcing period. Some cleaning action takes place during the arc period because the current is DCRP. The short circuit takes place anywhere between 20 and 200 times per second. This is controlled by the electrode wire speed control. The lighter the gauge metal to be welded, the slower the wire speed. The short arc method can be used to weld in all positions with excellent bead formation. The amount of heat injected in the weld zone is far less than the spray arc and the heat zone is smaller, which makes the short arc good for welding light

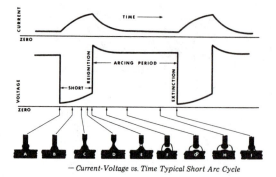

— *Current-Voltage vs. Time Typical Short Arc Cycle*

**Figure 31-4** The short arc cycle starts when the wire touches the weld pool at (A) and is completed at (I), then starts again. (*Courtesy of Linde Division of Union Carbide.*)

gauge material thereby reducing burn throughs. Carbon steels can be welded with short arc up to about 3/8″ (9.5 mm) thick.

## SPRAY ARC PROCESS

The spray arc differs from the short arc in the voltage and the amount of heat produced or injected into the base metal. In the spray arc, very small particles of molten metal from the electrode wire are rapidly propelled or transferred from the electrode wire (Figure 31-5) to the work piece through the arc stream. This is due to the high amount of current and voltage, which produces more heat, which is injected in the work piece. The spray arc operates on a DCRP power supply so that the electron movement is from the base metal to the wire electrode that is being fed into the molten pool. Unlike the TIG, more heat is liberated in the base metal than in the electrode due to the shorter distance between the wire electrode and the base metal. The metal ions, which are attracted to the base metal, are much hotter when they hit the work piece. As the wire melts into fine particles, the voltage causes the molten particles to fly off the electrode wire in a spray effect. The higher the voltage, the more force behind the spray, the finer the particles.

The effectiveness of the spray arc depends upon the distance between the electrode wire and the work piece with an average about 1/2″ (12.7 mm). If the distance becomes excessive, the arc stream will wander, become erratic, and the weld will be difficult to control. With the higher heat, larger electrode wire can be used resulting in deep weld penetration. The spray arc produces a directional force which is stronger than the effects of gravity. Because of the deep penetration of the weld, it is not recommended for light gauge metal or metal less than 3/32″ (2.4 mm) thick.

## SHIELDING GAS

Like the TIG welding process, the MIG process uses a shielding gas to displace the air in the weld zone to prevent contamination of the molten metal. Oxygen, nitrogen, and hydrogen, in the form of water vapor, must be kept out of the weld zone. If too much oxygen is in the weld zone, it combines with the carbon to form carbon monoxide, which can be trapped in the metal causing porosity. Also, oxygen combines with other elements in the steel to form compounds that could weaken the weld metal. Deoxidizers, such as manganese and silicon, will combine with the remaining oxygen and float to the top of the molten metal to form a slag. Using electric wire without a deoxidizer will lower the strength of the weld metal.

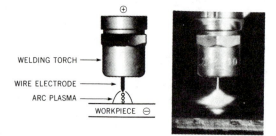

**Figure 31-5** Spray-arc welding. (*Courtesy of Linde Division of Union Carbide.*)

| SHORT ARC WELDING | | | |
|---|---|---|---|
| METAL | THICK-NESS | SHIELDING GAS | ADVANTAGES |
| | 14-gage or less | C-8 | Good burnthrough and distortion control. Shielding gas useful for spray arc welding. |
| CARBON STEEL | Less than 1/8" | 75% Argon-25% CO$_2$ (C-25) | High welding speeds without burnthrough. Minimum distortion and spatter. Best puddle control for out of position welding. Provides best mechanical properties for any given wire. |
| | Over 1/8" | 75% Argon-25% CO$_2$ (C-25) | High welding speeds without burnthrough. Minimum distortion and spatter. Best puddle control for out of position welding. Provides best mechanical properties for any given wire. |
| | | 50% Argon-50% CO$_2$ | Deep penetration, low spatter. |
| | | CO$_2$ | Deeper penetration. Faster welding speeds. |
| STAINLESS STEEL | | 75% Argon-25% CO$_2$ (C-25) | For use where corrosion resistance is not mandatory. |
| | | 90% Helium-7-1/2% Argon-2-1/2% CO$_2$ (A-1025) | No effect on corrosion resistance. Small heat affected zone. No undercutting; minimum distortion. Good bead shape and mechanical properties. |
| HIGH YIELD STRENGTH STEELS | | 60% Helium-35% Argon-5% CO$_2$ (A-415) | With LINDE 95, 105, or 120 wires, excellent arc stability, wetting characteristics, bead contour. Little spatter. High impacts. |

Figure 31-6  Short arc welding. (*Courtesy of Linde Division of Union Carbide.*)

Nitrogen in solidified steel reduces the ductility and impact strength of the weld metal and could cause cracking; if in excess, the weld metal would be porous. Hydrogen, if permitted to combine with the steel weld metal could cause .porosity and underbead cracking may occur.

There are three gases mainly used for MIG welding. They are argon, helium, and carbon dioxide. Of these three, argon and helium are inert. Carbon dioxide and oxygen are oxidizing agents and to compensate for these tendencies, a deoxidizer is used in the electrode wire formula. These gases can be used alone or in various combinations (Figures 31-6 and 31-7). For example, 75% argon and 25% carbon dioxide, or 95% argon and 5% oxygen.

Argon is seldom used by itself for welding steel, but it is used for welding aluminum, copper, and some other metals. When using only argon for steel, the welds have shallow penetration with a lack of fusion in the root area and poor bead formation and contour. With the addition of up to 5% oxygen, the bead contour and weld penetration is improved with reduced undercutting. Most of the argon-oxygen combinations are used for the spray arc process (Figure 31-8 and 31-9).

The combination of 75% argon and 25% carbon dioxide (CO$_2$), often referred to as C-25, produces less spatter, better arc stability, and smooth operation; it also makes it easier to bridge the gap on poor fit ups when used with the short arc process. There are many other combinations of argon-CO$_2$ such as 10% CO$_2$ or 50% CO$_2$. The mixture depends on the job requirement.

Helium, as mentioned in the TIG section, is less dense and much lighter than argon, which means that more helium in cubic feet per hour

| MIG WELDING-SPRAY ARC | | | |
|---|---|---|---|
| **METAL** | **THICK-NESS** | **SHIELDING GAS** | **ADVANTAGES** |
| ALUMINUM | 0-1/2" | Argon | Best metal transfer, arc stability and plate cleaning. Little or no spatter. |
| | 1/2-1" | 25% Argon 75% Helium | Higher heat input. Produces more fluid puddle and flatter bead. Minimizes porosity. |
| | 1/2" + | Helium | Highest heat input. Good for mechanized welding. |
| MAGNESIUM | | Argon | Excellent cleaning action. Provides more stable arc than helium-rich mixtures. |
| LOW ALLOY STEEL | | Argon-Oxygen (2%) | Reduces undercutting. Improves coalescence and bead contour. Good mechanical properties. |
| CARBON STEEL | | Argon-Oxygen (5%) Argon-$CO_2$ (C-8) | Improves droplet rate and arc stability. Produces a more fluid and controllable weld puddle; good coalescence and bead contour. Minimizes undercutting; permits higher speeds. |
| | | Argon-$CO_2$ (C-25) | Suitable for high current welding. |
| | | Argon-Oxygen-$CO_2$ | Improved results on mill scaled material when used with highly deoxidized wires. |
| STAINLESS STEEL | | Argon-Oxygen (1%) | Good arc stability. Produces a fluid and controllable weld puddle; good coalescence and bead contour. Minimizes undercutting. |
| | | Argon-Oxygen (2%) | Can be used on the more sluggish alloys to improve puddle fluidity, coalescence and bead contour. |
| COPPER, NICKEL, & Cu-Ni ALLOYS | 0-1/8" | Argon | Good arc stability. |
| | 1/8" + | Argon-Helium | Higher heat input of helium mixtures offsets high heat conductivity of heavier gages. |
| | 1/8" + | Helium | Higher heat input and improved penetration. |
| TITANIUM | | Argon | Good arc stability; minimum weld contamination. Inert gas backing is required to prevent air contamination on back of weld area. |

**Figure 31-7** MIG welding-spray arc: shielding gases for short and spray arc. (*Courtesy of Linde Division of Union Carbide.*)

or liters per minute is required than argon. Helium has a higher electrical resistance than argon, which makes argon more suitable for short arc welding because of the lower voltage required. Helium can be operated at a much higher voltage and is often used in automobile welding with greater welding speeds. By mixing argon and helium, the good characteristics of both gases are obtained, namely the penetration qualities of argon and the increased welding speed of helium. Most of the argon-helium combinations are used to weld non-ferrous metals such as aluminum, copper, nickel, and others.

NOTE: $CO_2$ cannot be used for the spray arc process because it turns the spray into a globular effect. Instead of the electrode coming off in fine particles, the end of the electrode builds up in the shape of a ball several times the electrode diameter before it is transferred to the work piece. Generally, the amount of $CO_2$

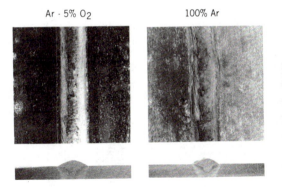

**Figure 31-8** Different combinations of shielding gases: effect of oxygen additions to argon. (*Courtesy of Linde Division of Union Carbide.*)

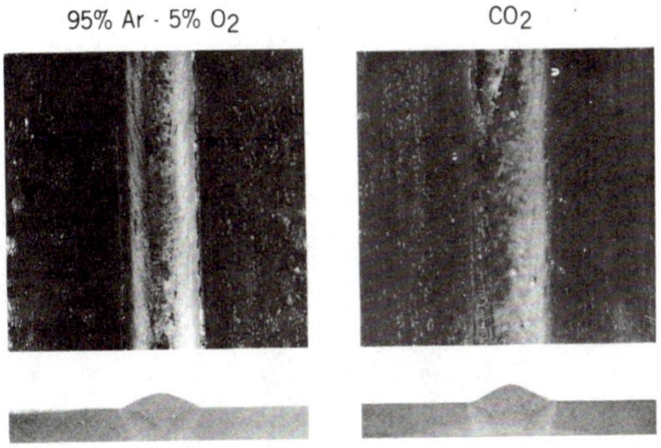

**Figure 31-9** Different combinations of shielding gases: Ar-5% $O_2$ and $CO_2$ shielding gas. (*Courtesy of Linde Division of Union Carbide.*)

used in combination with argon does not exceed 10%.

Carbon dioxide ($CO_2$) is commonly used for welding steel with the MIG short arc process (Figure 31-10). Carbon dioxide is made up of two elements, carbon and oxygen. During the welding operation when carbon dioxide is exposed to the high temperature of the electric arc, the carbon dioxide is changed into carbon monoxide and oxygen. This change increases the amount of oxygen in the weld zone, which could be damaging if the wrong type of electrode wire is used. When using $CO_2$ as a shielding gas, the electrode wire must contain deoxidizers that readily combine with the oxygen and prevent the oxygen from combining with the weld metal. Carbon dioxide is relatively inexpensive as compared to argon or argon-carbon dioxide combinations. Many of the undesirable characteristics of other shielding gases such as argon are eliminated. Carbon dioxide has a deep penetrating characteristic with a good bead contour and little tendency to undercut as compared to argon. There is more spatter in the weld zone, and the arc is more violent due to the electrical resistance of the gas as compared to the quieter arc and less spatter from argon-$CO_2$. The violent arc and spatter could lead to problems when welding thin

gauge metal or if overall appearance is important.

## ELECTRIC WIRE OR CONSUMABLE ELECTRODES

Selection of the proper electrode wire is very important in MIG welding. The electrode wire with the proper shielding gas determines the resulting physical and mechanical properties of the weld. Several factors to consider when selecting the electrode wire are: the work piece or base metal's chemical and mechanical prop-

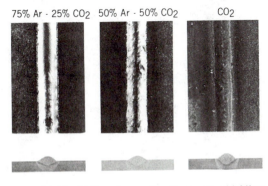

**Figure 31-10** Different combinations of shielding gases: effect of $CO_2$ additions to argon. (*Courtesy of Linde Division of Union Carbide.*)

# PLAIN CARBON STEEL WIRES

| APPLICATION | RECOMMENDED WIRE | FEATURES / BENEFITS | WELD MECHANICAL PROPERTIES |
|---|---|---|---|
| General purpose Structural steel work Automotive, railroad, farm and earthmoving equipment Ships and barges | LINDE 82 | One wire for many applications—*simplifies ordering; reduces inventory costs*<br><br>Good deoxidizing properties (contains silicon and manganese deoxidizers)—*good performance over rust and mill scale; produces sound quality weld metal using $CO_2$-containing shielding gases; low plate preparation costs* | Meets AWS ER70S-3 |
| Automotive components General purpose Structural | LINDE 29S | Lower cost wire with adequate deoxidizing properties Porosity-free welds over light mill scale | Meets AWS EM13K and ER70S-3 |
| Pipe Offshore drill rigs Structural steel work | LINDE 65 | Premium wire; contains Al, Zr and Ti deoxidizers *in addition* to silicon and manganese deoxidizers; minimal porosity; no holes, *superior quality welds even over heavy rust and mill scale and with all grades carbon steel; minimal plate preparation* | Meets AWS ER70S-2 |
| Pipe All-position welding | LINDE 85 | Excellent fluidity and wettability; very smooth, very flat bead on butt and fillet welds, regardless of position—*excellent weld appearance, upgraded product* | Meets AWS ER70S-4 |
|  | LINDE 85A | Same as LINDE 85 but contains even more manganese for still better wetter characteristics and an even flatter bead—*even better utilization of weld metal, lower weld metal cost*<br><br>Higher deoxidizer level, sounder welds in rimmed and semi-killed steels—*improved product quality* | Meets AWS ER70S-7 |
| Pipe High stress, such as truck drive shaft assemblies | LINDE 86 | Highest deoxidizing power of the plain carbon steel wires, no weld metal porosity; highest tensile strength (as welded) of the plain carbon steel wires, using $CO_2$ shielding gas—*superior quality welds*<br><br>Excellent wetting; smooth flat bead; maximum utilization of weld metal—*low weld metal cost* | Meets AWS ER70S-6 |

Figure 31-11 Typical electrode wire selection guide. (*Courtesy of Linde Division of Union Carbide.*)

spools from 1 to 60 pounds (6.9 to 414 kPa) in 4", 8", 12", and 14" (101, 203, 304, and 355 mm) spools (Figure 31-12), although for production use, larger spools or coils are available. The diameter of the electrode wire can be .030", .035", .045", 1/16", 5/64", and 3/32" (.8, .9, 1.1, 1.6, 2.0, and 2.4 mm). The .030" and .035" (.8 and .9 mm) diameters are generally used on thin-gauge metal or metal up to 3/16" (4.76 mm) in thickness.

Previously, deoxidizers were mentioned in the shielding gas section. Deoxidation is the combination of an element with oxygen from the weld puddle resulting in a slag or gloss formation on the surface (Figure 31-13). Removing the oxygen from the molten puddle eliminates it as a cause of porosity in the weld metal.

Silicon is the most commonly used deoxidizing element in electrode wire. Manganese is a deoxidizing agent, adds strength to the weld metal, and increases crack resistance. Aluminum, titanium, and zirconium (Figures 31-13 and 31-14) are also very strong deoxidizers, even when used in very small amounts. Carbon has a great effect on the structure and mechanical properties of the steel, more than any other element. The carbon content of the MIG electrode wire ranges between .05% to .12%. Increasing the carbon content in the wire or the work piece has an effect on porosity, particularly when using $CO_2$ as a shielding gas. By exceeding .12% of carbon, the electrode metal may lose some of the carbon content, causing possible porosity. Using deoxidizers will help to overcome this problem.

Other elements are added to the electrode wire content to improve mechanical and/or corrosion resistance properties. It is very important to match the electrode wire and the shielding gas with the work piece metal. When welding aluminum, stainless steel, and other metals, the electrode metal must be compatible with the work piece metal as well as the correct shielding gas.

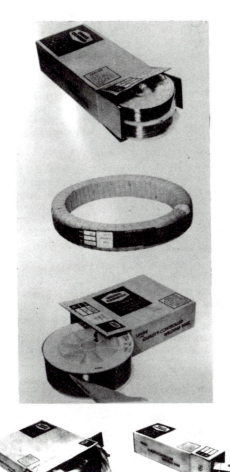

**Figure 31-12** MIG electrode wire available in different sizes of spools and coils. (*Courtesy of Linde Division of Union Carbide.*)

erties, the shielding gas to be used or available, the job requirements, and the type of joints to be welded (Figure 31-11).

The electrode wire is classified by the American Welding Society (AWS) as follows: using a E70S-2 wire for example, E identifies it as an electrode, the 70 is the tensile strength in psi per thousand (kPa), the S signifies a solid wire and 2 is the chemical composition of the electrode wire. Electrode wires are available in

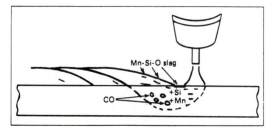

**Figure 31-13** Weld metal soundness. Sufficient deoxidation of the weld puddle is needed to minimize CO porosity in the solidified weld metal. Oxygen from the base metal, surface scale, shielding gas and welding wire combine with carbon in the weld to form CO gas porosity. Elements are added to the wire which can combine with the oxygen in preference to carbon to form harmless slags. These elements—called deoxidizers—are commonly manganese (Mn), silicon (Si), titanium (Ti), aluminum (Al), and zirconium (Zr). (*Courtesy of Linde Division of Union Carbide.*)

## MIG WELDING EQUIPMENT

The MIG welding equipment consists of a power unit suitable to supply DCRP current which can be either an AC rectified transformer or a DC generator, a wire feed unit and wire spool with either an enclosed or external wire spool, a shielding gas supply with a regulator-flowmeter, and a torch assembly with hoses either air or water cooled (Figures 31-15a and b).

There are many variations of the MIG welding equipment such as a wire feed (Figure 31-16) and torch assembly that can be connected to a suitable power supply (Figure 31-17). Another unit, made especially for welding light materials, has the wire feed and spool assembly enclosed in the welding machine cabinet. Several units of this type can also be used for AC and DC welding or can be used with an addon TIG assembly. This type of machine has an output of about 200 amperes (Figures 31-18 and 31-19).

For welding heavier material, the power supply must have a higher output, above 200 amperes. For average welding, 200 to 250 ampere machines are used. Constant or continuous welding on heavy material requires machines of 300 to 500 amperes with 100% duty cycle.

## CURRENT AND CONSTANT VOLTAGE

For MIG welding, the machine must be direct current (DCRP) with constant voltage (CV). Although the machine supplies a constant voltage to the arc, called arc voltage, the voltage will vary depending on the length of the arc. The length of the arc is the distance between the electrode and the work piece and can be changed by moving the gun farther away or closer to the work piece. The arc length (welding voltage) has an important effect on the type of process variation or the proper metal transfer. After the voltage has been set, machines with a voltage adjustment (some have not) can increase or decrease the arc gap; the machines will automatically increase or decrease the voltage. The change of voltage to the weld zone changes the burnoff rate of the electrode wire.

A lower voltage and higher wire speed will cause the wire to hit the work piece before it melts. Meltback or wire burning back to the contact tip is caused by high voltage and slow wire speed. Generally, the fine adjustments on welding voltage (arc length) can be made with the hand crank, and fine adjustments on welding current (wire feed speed) can be made with the wire feed control.

NOTE: In some areas, especially in some rural places, the MIG machine will react differently either day to day or throughout the day, due to the change in line voltage. Also, the machines may differ between a day shift and night shift. If the input voltage changes, the output voltage will change.

The short arc requires a lower voltage setting and wire speed as compared to the spray arc with high voltage and wire speed. On some of the power supply units or welding machines, they have what is called a slope control. Slope is the relationship between the amperage and the voltage output or the reduction in voltage output with the increasing current. Therefore, the constant voltage unit slope does not really provide constant voltage. The slope is used to limit the short circuit current to reduce the spatter. The greater the slope, the lower the current, the lower the spatter. With little or no slope in the circuit, the short circuit current

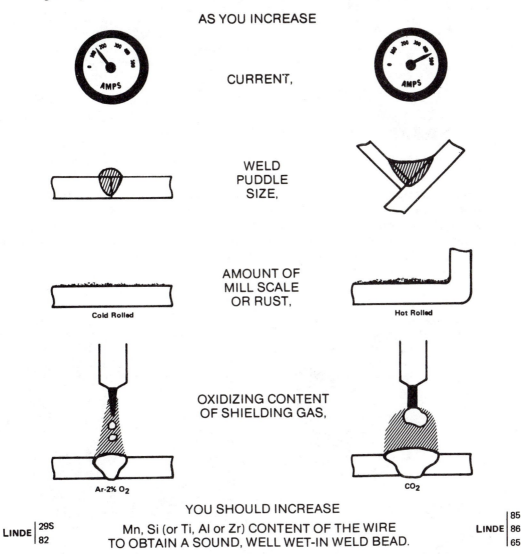

AS YOU INCREASE

CURRENT,

WELD
PUDDLE
SIZE,

AMOUNT OF
MILL SCALE
OR RUST,

Cold Rolled

Hot Rolled

OXIDIZING CONTENT
OF SHIELDING GAS,

Ar-2% O$_2$

CO$_2$

YOU SHOULD INCREASE

LINDE | 29S
82

Mn, Si (or Ti, Al or Zr) CONTENT OF THE WIRE
TO OBTAIN A SOUND, WELL WET-IN WELD BEAD.

LINDE | 85
86
65

Figure 31-14 How to make your wire selection. (*Courtesy of Linde Division of Union Carbide.*)

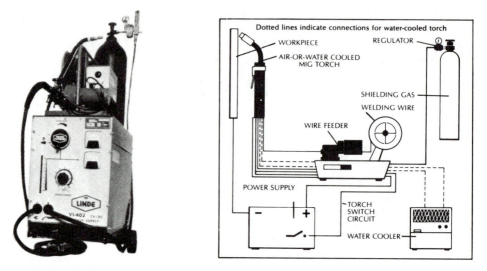

**Figure 31-15** Typical MIG welding set. (*Courtesy of Linde Division of Union Carbide.*)

goes too high causing a violent reaction resulting in spatter. Excessive slope (Figure 31-20) will cause the wire to carry full current and the short arc will not clear itself or burn off. It can cause the wire to pile up on the work piece. Different size or diameter welding cables, loose or dirty connection, and poor grounds are some of the things that will affect the slope, therefore the slope should be measured at the arc.

A new type of MIG welder was introduced recently, primarily to handle the new trends of thinner metal and frames in auto body repair. It

**Figure 31-16** Wire feed unit. (*Courtesy of Miller Electric.*)

**Figure 31-17** Constant potential rectifier type DC power source. (*Courtesy of Miller Electric.*)

**Figure 31-18** AC-DC arc welder and MIG combination. (*Photo by S. Suydam.*)

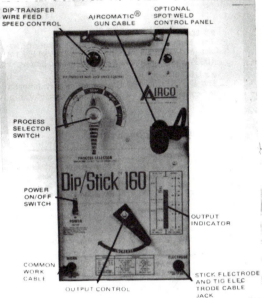

**Figure 31-19** Description of front panel in Fig. 31-15. (*Courtesy of Airco.*)

can be used also in sheetmetal shops for general light repairs or fabrications.

The electrode wire size ranges from .030" (.9 mm) to .035" (.89 mm ) for carbon or stainless steel, .035" (.89 mm) to 3/64" (1.191 mm) for aluminum, and to .068" (1.588) for flux core wire called Innershield.

The machine is capable of seam welding, spot welding, and a new feature called auto stitch mode.

The auto stitch mode gives repeated timed welds and timed off periods. Although the trigger is depressed throughout, the beads are made only in the on intervals. The length of the beads varies from a spot to beads about 1" (25.4 mm) long, during intervals from 1/2 to 2 seconds, with off periods of the same duration. The time intervals can be adjusted or changed from 1/2 to 2 second intervals. For example, to weld beads, depress the trigger and move along

the joint to be welded at a normal rate of travel. The beads will be deposited at intervals. Or, if small spots are required, a small bead or spot will be deposited, then there will be an off interval, then another small spot.

This method is useful for welding thin gauge metal because the off and on cycle tends to limit the heat buildup, thus reducing warping. Another feature is that the voltage control has five ranges, which gives better performance, especially in the lowest range.

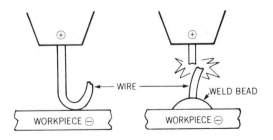

**Figure 31-20** Effect of too much slope. (*Courtesy of Linde Division of Union Carbide.*)

Although this type welding can be performed with a regular MIG gun, by depressing the trigger for an interval and releasing it, the new machine produces an evenly spaced bead with less chance of burn through on thin gauge metal. The auto stitch also can be used on heavier gauge metal.

## WIRE FEED UNITS

There are many styles of wire feed units. Some are enclosed in the welding machine cabinet, including the wire spool or reel (see Figure 31-18). These are generally lower amperage machines used for welding light gauge material. This type is very commonly used for repairing auto bodies where the wire spool or reel is protected from paint and grinding dust.

Other units are mounted on top of the welding machine or separated with a power input cable connected to the power source or welding machine. (see Figure 31-16). All of them have a wire feed that pulls the wire from the spool or reel and pushes the electrode wire through the torch cable. This type is used for hard wire, steel or stainless steel, and large diameter wire. Another type, the pull type, pulls the wire through the welding gun cable or contactor cable. The drive unit is located in the gun handle and is used for some aluminum and magnesium electrode wires which are relatively soft, although some harder aluminum wire can be used with the push type wire feed. The motor driven rollers of the wire feed units are made to accommodate one or more different diameter electrode wires or different rollers for different diameters.

NOTE: Using small diameter soft wire like 4043 aluminum on a push type wire feed is not very successful because the wire has a tendency to jam up at the rollers if pressure is put on the wire at the contact tip or the torch hose is sharply bent.

A control unit adjusts the wire speed. On the smaller MIG machines, the wire speed ranges

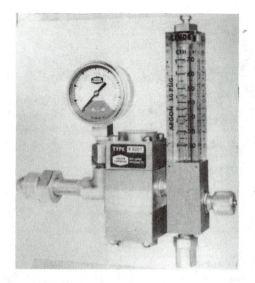

**Figure 31-21**  Typical regulator-flow meter. (*Courtesy of Linde Division of Union Carbide.*)

from 12″ to 20″ (304 mm to 508 mm) per minute for light gauge metal to approximately 290″ (7620 mm) per minute for welding up to 1/4″ to 5/16″ (6.35 mm to 7.94 mm) plates. The large machines have wire speeds up to about 900″ (22.86 mm) per minute. The spool or wire spindles are generally made to take different diameter spools (see Figure 31-12) or reels of wire.

**THE GAS SOLENOID**  The gas solenoid, which controls the flow of shielding gas, is mounted in the same unit. The shielding gas uses generally the same regulator and flowmeter (see Figure 31-21) as used for TIG welding except that the $CO_2$ cylinder outlet has a different type of connector. Also, the regulator-flowmeter is slightly different than that of argon or helium. Another feature of the $CO_2$ regulator-flowmeter is that the ones where there is a high consumption of $CO_2$ have a regulator-flowmeter with a built-in heater used to prevent the regulator from freezing and disrupting the flow of gas. When the trigger of the gun or a remote switch is depressed, the wire feeder mechanism and the gas solenoid are activated, as well as the energizing power supply contactor.

## WELDING GUN OR TORCH

The MIG welding gun (sometimes called a torch) assembly and hose assembly brings the electrode wire, shielding gas, and current from the wire feeder unit to the gun; the gun guides and directs the wire (Figure 31-22), shielding gas, and current to the weld area or zone. The guns are either air cooled or water cooled and are manually operated or semi-automatic. The air-cooled guns are used for lighter materials, if argon gas is used, and heavy metal if $CO_2$ shielding gas is used because of the cooling effect of the $CO_2$.

When used with argon or argon mixtures, the guns can weld light material using less than 150 amperes. The air-cooled guns with $CO_2$ shielding gas are rated as high as 500 amperes. Water-cooled guns are best for welding continuously at a higher ampere range with either argon or $CO_2$ shielding gas. The guns are available in either curved or straight front guide tubes for welding in various positions. At the end of the guide tube, there is a gas diffuser or collet where the contact tips are screwed into. An insulator or spatter guard slips over the diffuser, or collet. The nozzle is then fitted on the spatter guard or insulator. The nozzle directs the shielding gas to the weld area.

When using the MIG gun, the end of the nozzle must be checked for spatter build-up. An amount of build-up will hinder the flow of shielding gas; if there is too much build-up, it can come in contact with the electrode wire. The nozzle should be checked frequently. An anti-spatter spray can be used to slow down the spatter build-up.

The contact tips are available for different diameter wires. The contacts should be replaced if the diameter of the hole gets too large or egg-shaped. Occasionally, the electrode wire will fuse together with the contact tip due to excessive spatter build-up or holding the gun too close to the work. Normally, most MIG operators have several new contact tips available by the machine because the tips wear out readily through constant use and high heat.

**SPOOL GUN** Another type MIG gun or torch is a combination of a MIG gun and wire spool assembly sometimes called a spool gun (Figure 31-23). The gun assembly consists of a wire feed motor and drive rollers, a spool of consumable electrode wire, guide tube and liner with contact tip, a wire feed control, and a length of power cable and gas hose. The wire electrode weighs about 4 lbs. (1.81 kg) for steel and about 1 lb. (.45 kg) for aluminum, and is 4" (101 mm) in diameter.

Because the gun and wire are combined into one unit, the gun can be used a long distance

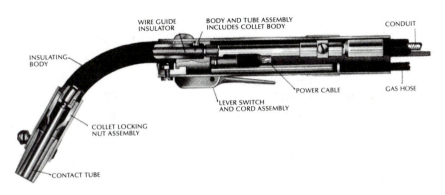

**Figure 31-22** Air-cooled MIG torch. (*Courtesy of Linde Division of Union Carbide.*)

**Figure 31-23** Using a MIG (GMAW) spool gun or torch. (*Courtesy of Linde Division of Union Carbide.*)

from the power supply up to 150 ft. (45.7 meters) from the source. The electrode wire diameter ranges from .030" to .045" (.9 mm to 1.1 mm) in steel wire and from .030" to 1/16" (.9 mm to 2.0 mm) in aluminum. The wire speeds range from 40 to 900 inches per minute (1 to about 20.5 mm) with some guns having different motor ratios for different ranges of wire feed.

guns have the contact tip extending slightly beyond the nozzle. The actual stickout or wire extension of the electrode wire and the distance between the work piece and the gun will vary with different manufacturer's recommendations. The actual stickout for guns used on eight gauge material is about 1/4" to 3/8" (6.35 mm to 9.5 mm); for heavier material, it is about 3/8" to 1/2" (9.5 mm to 12.7 mm).

The amount of stickout has an influence on the current. The stickout, or wire extension, is preheated, and the greater the stickout, the greater the preheat. This requires less current to melt the wire at a given wire speed. The machine will automatically reduce the amount of current output. By decreasing the amount of stickout and preheating, the current will be increased in order to melt the wire. Holding the gun too far away or increasing the work distance will also generally increase the stickout distance and increase the amount of deposited electrode wire at lower welding heat. This results in decreased penetration and poor bead formation. Reducing the work distance and stickout can cause the wire to burn back and fuse with the contact tip.

For lighter duty guns using .03" to .035" (.76 mm to 89 mm) wire, the work distance should be about 1/2" (12.7 mm); for heavy-duty guns, about 3/4" (19.0 mm) or depending on the manufacturer's recommendations. Welding fillet or inside corner welds, the work dis-

## WIRE STICKOUT OR WORK DISTANCE

The wire stickout is the distance between the end of the contact tip to the end of the electrode wire. The work distance is the distance between the end of the nozzle and the work piece (Figure 31-24). How to determine the work distance varies slightly with different MIG guns. On some guns, the contact tip is slightly inside the edge of the nozzle; other

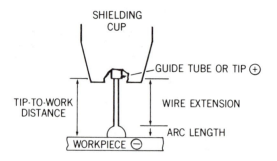

**Figure 31-24** Tip-to-work distance.

tance may have to be increased if the nozzle opening is large which will interfere with getting close to the weld area.

## CURRENT SELECTION

Current selection is very important to get the correct penetration, and to avoid cold lapping, undercutting, and excessive penetration or burn through. The MIG offers a wide selection of welding currents and thicknesses of metals by using the same diameter electrode wire. For example, a MIG welder for light gauge material can weld from 20 gauge, .037" (.91 mm) to 3/16" (4.76 mm) or up to 1/4" (6.35 mm) depending on the type of joint. Likewise, different thicknesses of metal can be welded without difficulty.

As the thickness of the metal increases, the current and the voltage increase. For short arc, as the thickness of the metal increases, the current, voltage and wire speed also increase. Using .035" (8.9 mm) wire to weld 1/16" (1.67 mm) steel, the current could range from 80 to 110 amperage, the voltage 17 to 20, and the wire speed could reach about 180 inches per minute. Using 3/16" (4.7 mm) the amperage is about 140 to 160, with 20 to 24 volts, and a wire speed of almost 245 IPM.

Although many charts on welding currents and thicknesses of metals to be welded are available, the operator must determine the correct current either by experience or by trial. Sometimes, having several charts lead to confusion; for example, one chart specifies about 150 amps for .035" (89 mm) electrode wire, and another one recommends 300 amps. If time permits, using scrap of the same thickness, set the machine to an approximate current, run a bead, and observe the results.

Another factor to consider is that one machine may appear to be better or seem to

**Figure 31-25** Direction of arc travel, forehand welding technique. Longitudinal torch positions. (*Courtesy of Linde Division of Union Carbide.*)

have better penetration than another machine set at the same amperage.

## TRAVEL

### SPEED OF TRAVEL

Speed of travel is very important because it has a definite effect on penetration. When the gun is moved too fast, the penetration of the work piece will be too shallow and the bead too narrow and poorly formed. Welding speeds that are too slow will cause excessive heat in the weld zone, excessive build-up of weld metal, possible burn or metal burn through, and perhaps cold lap or metal not fused to the work piece.

### METHODS OF TRAVEL

There are two methods of traveling or running beads with the MIG gun, either the pushing or the pulling method. The pushing, or leading, method has the gun pointed at about a 5° to 15° angle in the direction of travel. The pulling, or dragging, method has the gun pointed away from the direction of travel.

The pushing (or leading) method, sometimes called forehand welding (Figure 31-25), is used on heavier materials with greater welding speed.

**Figure 31-26** Direction of arc travel, backhand welding technique. Longitudinal torch positions. (*Courtesy of Linde Division of Union Carbide.*)

The pulling (or dragging) method, sometimes called backhand (Figure 31-26) is generally used on light gauge material. The pulling method gives the operator a better view of the bead formation, while the bead is partially covered by the pushing or leading method. As an individual becomes more experienced with MIG welding, he may adopt a method which is best suited for the job at hand.

The angle which the gun is tilted or pointed either forward or backward in the direction of travel is called the longitudinal gun (Figures 31-25 and 31-26) or torch angle. The angle in which the gun or torch is held in relation to the work piece is called the transverse torch angle. For example, when welding a bead on a flat plate, the welder holds the gun at nearly a 90° angle from the horizontal (Figure 31-27) or

vertical and points it forward or backward at a 5° to 15° angle. While welding an inside corner or fillet weld, the welder holds the gun at a 45° angle from the horizontal or vertical (Figure 31-28).

The transverse angles are very similar to the angles used when welding with stick electrodes. The major problem with MIG welding is tipping or tilting the gun at an excessive angle, which will affect the shielding gas curtain, which can contaminate the weld metal and cause porosity. This is sometimes the case when repairing equipment or fabricating with MIG and due to obstruction, the gun cannot be held at the proper angle.

## WELDING PROCEDURES

### PREPARATION

Like the TIG welding machine, the MIG must be thoroughly checked out before attempting to weld. All cable and hose connections should be checked for tightness and condition. Never assume that these were checked the day before. Look for worn spots, burn marks, or possible defects, especially if the wire feed unit is separated from the power supply. Check the contact tip for wear or spatter deposit; replace if necessary. A worn contact can affect the wire speed or cause it to fluc-

**Figure 31-27** Transverse direction. Transverse torch positions. (*Courtesy of Linde Division of Union Carbide.*)

**Figure 31-28** Transverse direction. Transverse torch positions. (*Courtesy of Linde Division of Union Carbide.*)

tuate. Check the nozzle for spatter; clean it, if it is dirty. Do not strike the nozzle against hard objects to remove spatter. The opening can be distorted, which will alter the shielding gas curtain. Also, it may crack the insulator, if it has one.

Depending on the type of machine being used, set the wire feed in relation to the metal being welded. Set ampere and voltage control, if applicable, to the machine or unit. Some MIG units indicate the wire speed and amperes in relation to the thickness of metal to be welded. Some recommend a voltage setting of about 20 volts, if adjustable. Open the shielding gas cylinder valve. Turn the welding machine on, depress the gun lever or button to activate the shielding gas solenoid, and set the gas flow—which can range from 10 to 50 CFH (5 to 23 LPM) depending on the thickness of the metal, type of shielding gas, and the type of metal. Check the water flow if the gun is water cooled.

With the machine on, observe the wire feed speed and check for steady operation. If the wire feed is separate, turn on the wire feed. Turn off the machine and cut the electrode wire to the proper stickout. Select a steel flat plate or sheet approximately 4" X 6" (101 mm X 152 mm) and about 3/16" to 1/4" (4.76 mm to 6.35 mm) thick. Prepare the plate or work

piece by removing rust, scale, paint, or dirt. If it is aluminum, remove the oxide coating from the surface. Securely attach the ground clamp.

NOTE: Be sure to check the surrounding area for flammable material. It is too late after a fire has started.

## METHODS OF STARTING THE ARC

There are several different methods that can be used to start the arc. Two of the most commonly used methods are the scratch method and the run-in method.

The scratch method (Figure 31-29) is performed by scratching the end of the electrode wire or the work piece, just ahead of the weld area or zone, at the same time the lever or button of the gun is depressed to start the flow of current, shielding gas, and wire. Sometimes the arc is started on a tab or metal strip next to the weld area so that the arc is well established as it reaches the weld area. These are sometimes called run-on tabs but cannot be used for all welding operations. They are used more frequently on aluminum and materials other than steel.

The run-in method (Figure 31-30) involves holding the gun about 1/2" to 3/4" (12.7 mm to 19 mm), depending on length of stickout, above the surface of the work piece. When the lever or button is depressed activating the current, shielding gas, and wire feed, the electrode wire will contact the surface and establish the arc.

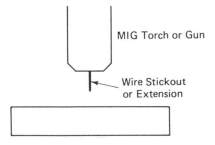

MIG Torch or Gun

Wire Stickout or Extension

**Figure 31-29** Scratch method to start arc. Scratch end of electrode wire just ahead of weld area or on an adjoining tab. At the same time, depress trigger to start arc.

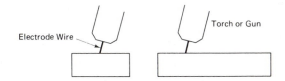

Electrode Wire

Torch or Gun

**Figure 31-30** Run on method. Hold torch or gun above surface with electrode about 1/2" (12.7 mm) from work piece. Depress trigger to start electrode; as electrode touches work piece, the arc will start.

In either method, the surface of the work piece or run-on tab must be clean or free from rust, scale, paint, or dirt in order to make the proper contact to start the arc. Another method suggested by a manufacturer is to run out about 2" (50.8 mm) of electrode wire (Figure 31-31), hold the gun about 1/2" to 3/4" (12.7 mm to 19 mm) above the surface of the work piece, bend the wire for a good contact, then depress the lever or button to start the arc. The remaining piece of wire is broken off after completing the bead. In any starting method, after the arc is established, raise or lower the gun to the recommended distance from the work piece with the proper gun angle.

## STARTING THE WELD

NOTE: After the different welding processes discussed previously, MIG welding is one of the easiest of the welding processes to learn in regard to the skill required.

Like the other types of welding, the new operator should practice in flat plate to get used to the gun and its operation. With the machine properly adjusted to the thickness of the metal, strike an arc at one end of the plate and run a bead the length of the plate. Hold the gun at right angles to the work piece, called the transverse angle, and tilt the gun forward or backward about 5° to 15° in the direction of travel. Do not tilt the gun too far because it

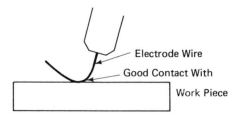

Figure 31-31 With wire extended out about 2" (50.8 mm) and in contact with work piece, depress work trigger to start arc.

— Electrode Wire

— Good Contact With

Work Piece

may distort the gun shield. Before breaking the arc, reverse the direction of travel at the end of the bead, then move backward across the end of the bead with increased speed, and release the trigger. This will help to eliminate the crater at the end of the bead. Check the bead for appearance, width, and contour. It may be necessary to increase or decrease the current setting or the wire speed to obtain the proper bead. Continue running beads, starting, and stopping in the middle. Try using different methods of striking the arc. Some prefer one to the other. Also, practice using pulling and pushing methods of travel in practice.

NOTE: Lighter gauge metal can be used for practice; the thickness mentioned is only a suggestion.

## JOINT PREPARATION

Joint preparation for MIG is basically the same as for stick electrodes with some exceptions. Due to the deep penetration of the MIG process, the root opening and root face can be narrower or smaller, which means the total angle of the joint opening can be less.

# JOINTS

## BUTT JOINTS

Using 2" × 6" (50.8 mm × 155 mm) plate, 3/16" to 1/4" (4.76 mm to 6.35 mm) thick, tack weld the two plates together leaving a root opening approximately one-half the thickness of the plates. No joint preparation or bevelling is necessary if the plates are less than 1/4" (6.35 mm) thick, but for quality welds, some bevelling may be necessary.

Plates over 1/4" (6.35 mm) should be bevelled to about a 60° total angle. Hold the gun directly over the center of the joint using the correct wire stickout angle with about 3/8"

## FLAT POSITION

A. SINGLE PASS BUTT WELD
 — LONGITUDINAL TORCH ANGLE = 5° - 10°
 — TRANSVERSE  TORCH ANGLE = 90°

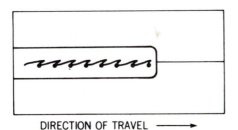

DIRECTION OF TRAVEL ⟶

**Figure 31-32** A method of torch manipulation for butt welding in a flat position. (*Courtesy of Linde Division of Union Carbide.*)

(9.5 mm) wire stickout (see Figures 31-27 and 31-32). Use the pushing method on half the joint and pulling on the other half, ending each bead in a backwards movement. Check the penetration, size of bead, and appearance. If preferred, weld the other side of the joint; slightly change the settings and note the difference. By carefully checking the beads, mistakes can be corrected.

Practice butt welds using two different thicknesses of metal. In this case, the transverse angle should be decreased slightly so the gun is pointed more toward the heavier piece of metal to prevent the lighter or thinner piece from burning through.

## LAP JOINTS

Place two pieces of metal together to form a lap joint, using metal with the same thickness and dimensions as used in the butt joint. The metal can be either clamped or tack welded together. Hold the gun at a 45° angle from the horizontal plate (see Figure 31-28) with about a 3/8″ to 1/2″ (9.5 mm to 12.7 mm) wire stickout. Start the arc at one end and use a pulling method of travel with the gun tilted at about a 5° to 15° angle. Turn the plates over and weld

the other lap joint using a pushing method of travel. Examine the bead appearance and penetration. Adjust the machine if necessary to obtain a better bead. If undercutting is present on the bottom plate or melting away of the top plate edge, adjust the transverse angle. Try some more joints using different thicknesses of metal. These types of joints are common in transportation vehicles when lighter gauge metal pieces are welded to a heavier gauge such as frame. Try to avoid burning through the light gauge metal.

## FILLET OR INSIDE CORNER WELDS

Using two pieces of metal, position one piece vertically in the center of the horizontal piece and tack weld the ends to hold in position. Because the nozzle interferes with bringing the gun in close to the weld area, use a greater wire stickout, up to 3/4″ (19 mm). Hold the gun so it bisects the angle of the two plates or at a 45° angle (Figure 31-33). Start the arc at one end, using a pulling method of travel with the gun at a 5° to 15° angle and weld the length of the joint. Examine the bead for appearance and penetration. If the vertical plate leans due to distortion, weld the opposite side of the plate.

Additional joints can be welded, but lean the vertical plate outward, then tack weld on

A. FILLET WELD
 — LONGITUDINAL TORCH ANGLE = 5° - 10°
 — TRANSVERSE  TORCH ANGLE = 45°

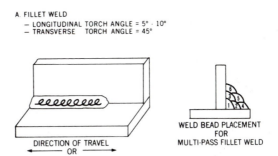

DIRECTION OF TRAVEL
— OR —

WELD BEAD PLACEMENT
FOR
MULTI-PASS FILLET WELD

**Figure 31-33** Torch manipulation for fillet or inside corner welds. (*Courtesy of Linde Division of Union Carbide.*)

the ends, similar to welding fillet welds using a stick electrode. Or use a staggered or intermittent welding procedure; that is, make a short bead on one side of the vertical plate then make a short bead on the opposite side to prevent distortion. If undercutting is present, change the transverse angle of the gun.

## OUT-OF-POSITION WELDING

Both spray arc and short arc can be used for out-of-position welding, such as horizontal (Figure 31-34), overhead (Figure 31-35) and vertical up or down (Figure 31-36). Spray arc with its strong directional force is not good for out-of-position welding.

The short arc can be used for out-of-position welding such as horizontal, overhead and vertical. Spray arc is limited to flat or horizontal welding, except when using aluminum, with its high cooling ratio, to freeze the weld metal. Although the penetration is relatively good, the high quantity of fluid or molten metal in the puddle makes it unsuitable for vertical and overhead welding.

With short arc in out-of-position welding, it may be necessary to reduce the current slightly to avoid having too much molten metal in the weld area at one time.

OVERHEAD POSITION

– LONGITUDINAL TORCH ANGLE = 5° - 10°
– TRANSVERSE TORCH ANGLE = 90° ✱

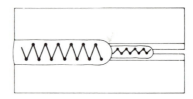

– USE SAME WEAVE TECHNIQUE FOR FILLET WELDS
– SINGLE PASS WELDS USE SAME WEAVE TECHNIQUE AS FIRST PASS SHOWN

PAUSE AT ●

✱REMEMBER TO ANGLE TORCH TOWARD WALLS WHEN WEAVING

**Figure 31-35** Torch manipulation for overhead welding. (*Courtesy of Linde Division of Union Carbide.*)

NOTE: A rather new development called Pulsed Arc by Airco Welding makes it possible to use spray arc in all positions. In Pulsed Arc welding, the welding current switches from a low level to a higher level and back again automatically. The spray action only takes place at the higher level, called pulsed peak current.

## MULTI-PASS WELDING

Heavier metal can be welded by using two or more passes or beads following the same pattern or sequence of passes as used in stick electrode (Figure 31-37).

## WELDING MIG WITH ALUMINUM

Aluminum welding with MIG has the advantage over some of the other welding processes because it is faster and heavier pieces can be welded without difficulty. Argon shielding gas is used in place of $CO_2$ or argon-$CO_2$ for weld-

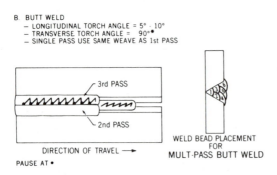

B. BUTT WELD
   – LONGITUDINAL TORCH ANGLE = 5° - 10°
   – TRANSVERSE TORCH ANGLE = 90° ✱
   – SINGLE PASS USE SAME WEAVE AS 1st PASS

3rd PASS
2nd PASS

DIRECTION OF TRAVEL ⟶
PAUSE AT ●

WELD BEAD PLACEMENT FOR MULT-PASS BUTT WELD

✱ REMEMBER TO ANGLE TORCH TOWARD WALLS WHEN WEAVING

**Figure 31-34** Torch manipulation for horizontal welding. (*Courtesy of Linde Division of Union Carbide.*)

VERTICAL POSITION

A. VERTICAL UP TRAVEL
— LONGITUDINAL TORCH ANGLE = 10° - 15°
— TRANSVERSE   TORCH ANGLE = 90°*

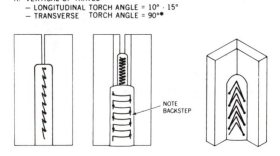

NOTE
BACKSTEP

B. VERTICAL DOWN TRAVEL
— LONGITUDINAL TORCH ANGLE = 5° - 10°
— TRANSVERSE   TORCH ANGLE = 90°*
— USE SAME WEAVE TECHNIQUE FOR FILLET WELDS
— SINGLE PASS WELDS USE SAME WEAVE TECHNIQUE
  AS FIRST PASS SHOWN

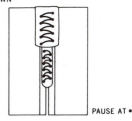

PAUSE AT •

\* REMEMBER TO ANGLE TORCH TOWARD WALLS WHEN WEAVING

**Figure 31-36A** Torch manipulation for vertical welding. (*Courtesy of Linde Division of Union Carbide.*)

VERTICAL UP          VERTICAL DOWN

DIRECTION OF ARC TRAVEL

DIRECTION OF ARC TRAVEL

**Figure 31-36B** Welding in the vertical position, up travel and down travel. (*Courtesy of Linde Division of Union Carbide.*)

ing aluminum up to about 1″ (25.4 mm) in thickness. For heavier thickness, argon and helium are used in combination. Argon gas produces less spatter and a stable arc.

Higher current output welders are used for welding heavy aluminum due to aluminum heat conductivity. For welding thinner aluminum, the current setting is slightly higher than welding steel. Spray arc is used on heavier material and is not recommended on material less than about 1/4″ (6.35 mm) thick. The short arc is recommended for lighter gauge material. Because of the colder arc, it can be used for out-of-position welding. Welding out-of-position is more difficult with the spray arc because of the high heat input.

Generally, when the MIG is used for welding heavier material, a water-cooled gun is used due to the high current. Air-cooled guns are used for welding lighter material.

## JOINT PREPARATION

On light gauge aluminum up to about 5/16″ (7.9 mm) in thickness, it requires little or no preparation for butt welds because of the deep penetration. Joints involving heavier materials are prepared very similar to steel but with narrow openings. However, there is a danger of melt through unless the current is reduced.

MULTIPASS BUTT WELD
— LONGITUDINAL TORCH ANGLE = 5° - 10°
— TRANSVERSE   TORCH ANGLE = (SEE BELOW FOR WEAVING
                               TECHNIQUE)

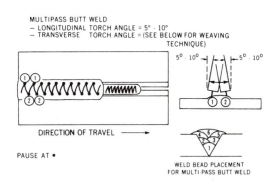

5° - 10°        5° - 10°

DIRECTION OF TRAVEL

PAUSE AT •

WELD BEAD PLACEMENT
FOR MULTI-PASS BUTT WELD

**Figure 31-37** Torch manipulation for welding heavy plate. (*Courtesy of Linde Division of Union Carbide.*)

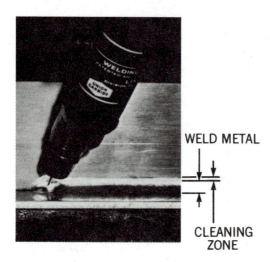

**Figure 31-38** Aluminum welding technique.

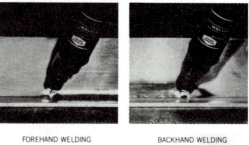

FOREHAND WELDING
TECHNIQUE

BACKHAND WELDING
TECHNIQUE

**Figure 31-39** Effect of welding technique—aluminum.

NOTE: With all aluminum, the surface to be welded must be prepared by removing the oxide formation. Unlike steel welding with MIG, only the forehand and vertical up welding techniques can be used. This is necessary to provide "cleaning" of the work piece in front of the weld puddle. Cleaning action is shown in Figure 31-38. The aluminum oxide skin on the metal is broken up by the arc action. If the background or vertical down welding technique were used, the bead may be porous, discolored, and exhibit poor wetting to the base plate. A comparison of welds made with the same welding conditions but using the backhand and forehand techniques is shown in Figure 31-39.

## ELECTRODE WIRE

Some types of electrode wire, such as .035″ (.9 mm) 4043, are more difficult to use than heavy electrode wire because there is a tendency of the wire to jam up at the wire feed, especially if the cable liner is worn. This is very common with the smaller short arc MIG welding machine. Generally, a harder wire can be used with little problem, unless the MIG gun has a pull unit in the gun. This is one of the many advantages of using a spool in the gun for small diameter wire. The composition of the electrode is similar to that used for TIG welding.

## FLUX CORE OR TUBULAR WIRE WELDING

Flux core or tubular wire welding (FCAW) is becoming increasingly popular as smaller diameter tubular type electrode wires are being developed.

Flux core is known by several different trade names, such as Airco Fluxcor, Hobart Fabco, and others. The electrode wire, instead of being solid, is tubular and the center part is filled with a flux. The electrode wire is designated by AWS as E70T-2, with the T designating tubular wire. The common types are E70T-2 to E70T-4, which are similar to the solid electrode wire.

The flux in the electrode provides deoxidizers, scavengers, arc stabilizers, slag formers, and alloying agents. The slag produced shields the weld metal and slows the cooling of the metal. A shielding gas of $CO_2$ is used to shield the weld puddle along with those gases produced by the flux.

The tubular wire produces a deeper, yet narrower, bead and less joint preparation than welding with stick electrodes.

| FAULT OR DEFECT | CAUSE AND/OR CORRECTIVE ACTION |
|---|---|
| 1) POROSITY | A. OIL, HEAVY RUST, SCALE, ETC. ON PLATE<br>B. WIRE – MAY NEED WIRE HIGHER IN Mn AND Si<br>C. SHIELDING PROBLEM; WIND, CLOGGED OR SMALL NOZZLE, DAMAGED GAS HOSE, EXCESSIVE GAS FLOW, ETC.<br>D. FAILURE TO REMOVE GLASS BETWEEN WELD PASSES<br>E. WELDING OVER SLAG FROM COVERED ELECTRODE |
| 2) LACK OF PENETRATION | A. WELD JOINT TOO NARROW<br>B. WELDING CURRENT TOO LOW; TOO MUCH ELECTRODE STICKOUT<br>C. WELD PUDDLE ROLLING IN FRONT OF THE ARC |
| 3) LACK OF FUSION | A. WELDING VOLTAGE AND/OR CURRENT TOO LOW<br>B. WRONG POLARITY, SHOULD BE DCRP<br>C. TRAVEL SPEED TOO LOW<br>D. WELDING OVER CONVEX BEAD<br>E. TORCH OSCILLATION TOO WIDE OR TOO NARROW<br>F. EXCESSIVE OXIDE ON PLATE |
| 4) UNDERCUTTING | A. TRAVEL SPEED TOO HIGH<br>B. WELDING VOLTAGE TOO HIGH<br>C. EXCESSIVE WELDING CURRENTS<br>D. INSUFFICIENT DWELL AT EDGE OF WELD BEAD |
| 5) CRACKING | A. INCORRECT WIRE CHEMISTRY<br>B. WELD BEAD TOO SMALL<br>C. POOR QUALITY OF MATERIAL BEING WELDED |
| 6) UNSTABLE ARC | A. CHECK GAS SHIELDING<br>B. CHECK WIRE FEED SYSTEM |
| 7) POOR WELD STARTS OR WIRE STUBBING | A. WELDING VOLTAGE TOO LOW<br>B. INDUCTANCE OR SLOPE TOO HIGH<br>C. WIRE EXTENSION TOO LONG<br>D. CLEAN GLASS OR OXIDE FROM PLATE |
| 8) EXCESSIVE SPATTER | A. USE $Ar\text{-}CO_2$ OR $Ar\text{-}O_2$ INSTEAD OF $CO_2$<br>B. DECREASE PERCENTAGE OF $H_e$<br>C. ARC VOLTAGE TOO LOW<br>D. RAISE INDUCTANCE AND/OR SLOPE |
| 9) BURNTHROUGH | A. WELDING CURRENT TOO HIGH<br>B. TRAVEL SPEED TOO LOW<br>C. DECREASE WIDTH OF ROOT OPENING<br>D. USE $Ar\text{-}CO_2$ OR $Ar\text{-}O_2$ INSTEAD OF $CO_2$ |
| 10) CONVEX BEAD | A. WELDING VOLTAGE AND/OR CURRENT TOO LOW<br>B. EXCESSIVE ELECTRODE EXTENSION<br>C. INCREASE INDUCTANCE<br>D. WRONG POLARITY, SHOULD BE DCRP<br>E. WELD JOINT TOO NARROW |

Figure 31-40 Weld defects—their causes and how to correct them. (*Courtesy of Linde Division of Union Carbide.*)

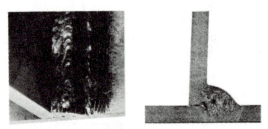

—*Examples of Porosity*

—*Examples of Lack of Penetration*

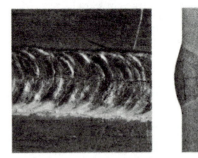

—*Example of Longitudinal Cracking*

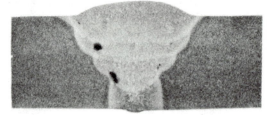

—*Example of Lack of Fusion*

—*Examples of Undercutting*

**Figure 31-41** Some of the welding defects. (*Courtesy of Linde Division of Union Carbide.*)

## QUESTIONS

1. What type electrode is used in MIG or gas metal arc welding (GMAW)?
2. What are some of the advantages of MIG?
3. What metals are commonly welded with MIG?
4. What are some of the trade names for the MIG process?
5. What are the different types of MIG processes?
6. In the short arc process, when is the metal transferred or deposited?
7. In what form is the metal transferred?
8. In the spray arc process, in what form is the metal transferred?
9. Which process is best for thin gauge metal, the short arc or spray arc?

10. What elements must be kept out of the weld zone to avoid contamination?

11. What elements or compounds are added to the electrode wire to help remove the remaining oxygen?

12. What are some of the shielding gases used for MIG welding?

13. What are some of the factors that must be considered when selecting electrode wire?

14. What type current is used for MIG welding?

15. What causes the wire to hit the work piece before it melts?

16. What is slope in reference to MIG welding?

17. What are some of the things that could affect the slope?

18. Why is it necessary to check the nozzle frequently?

19. What is a spool gun?

20. What is wire stickout?

21. Why is the correct current setting important?

22. What are some of the defects caused by the speed of travel being too fast?

23. What are the two methods of travel?

24. What are the two commonly used methods for starting the arc?

25. Is the current setting for steel and aluminum the same for the same thickness of metal?

26. What is the problem caused by using soft aluminum wire in some MIG machines?

# APPENDIX

# NOMINAL WRENCH OPENINGS TO FIT STANDARD BOLTS, NUTS AND SCREWS

## THREAD DIAMETER OF BOLT, NUT OR SCREW

Determine the type of bolt, nut or screw and locate the thread diameter below. Then move across to the left hand column to find the wrench size that will fit the bolt head or nut.

| NOMINAL WRENCH SIZE — Also width across flats of Bolts, Screw Heads and Nuts | AMERICAN STANDARD ASSOCIATION (B18.2.1 and B18.2.2—1965) | | | | | OLD U.S. STANDARD | | OLD S.A.E. STANDARD |
| | BOLTS | | NUTS | | | | | |
| | Square Bolt, Hex Bolt, Hex Cap Screw (Finished Hex Bolt), Lag Screw | Heavy Hex Bolt, Heavy Hex Screw, Heavy Hex Structural Bolt | Square Nut | Hex Flat, Hex Flat Jam, Hex, Hex Jam, Hex Slotted, Hex Thick, Hex Thick Slotted, Hex Castle | Heavy Square, Heavy Hex Flat, Heavy Hex Flat Jam, Heavy Hex, Heavy Hex Jam, Heavy Hex Slotted | National Coarse Bolts and Nuts | Cap Screws | National Fine Bolts, Nuts & Screws |
|---|---|---|---|---|---|---|---|---|
| 9/32" | No. 10* | — | — | — | — | — | — | — |
| 5/16" | — | — | — | — | — | — | — | — |
| 11/32" | — | — | — | — | — | — | — | — |
| 3/8" | 1/4"* | — | — | — | — | — | — | — |
| 7/16" | 1/4" | — | 1/4" | 1/4" | — | — | 1/4" | 1/4" |
| 1/2" | 5/16" | — | — | 5/16" | 1/4" | 1/4" | 5/16" | 5/16" |
| 9/16" | 3/8" | — | 5/16" | 3/8" | 5/16" | — | 3/8" | 3/8" |
| 19/32" | — | — | — | — | — | 5/16" | — | — |
| 5/8" | 7/16" | — | 3/8" | — | — | — | 7/16" | 7/16" |
| 11/16" | — | — | — | 7/16" | 3/8" | 3/8" | — | — |
| 3/4" | 1/2" | — | 7/16" | 1/2" | 7/16" | — | 1/2" | 1/2" |
| 3/4" | — | — | — | — | — | 7/16" | — | — |
| 13/16" | 9/16" | — | 1/2" | — | — | — | 9/16" | — |
| 7/8" | — | 1/2" | — | 9/16" | 1/2" | 1/2" | 5/8" | 9/16" |
| 15/16" | 5/8" | — | — | 5/8" | 9/16" | — | — | 5/8" |
| 31/32" | — | — | — | — | — | 9/16" | — | — |
| 1" | — | — | 5/8" | — | — | — | 3/4" | — |
| 1 1/16" | — | 5/8" | — | — | 5/8" | 5/8" | — | 3/4" |
| 1 1/8" | 3/4" | — | 3/4" | 3/4" | — | — | 7/8" | — |
| 1 1/4" | — | 3/4" | — | — | 3/4" | 3/4" | 1" | 7/8" |
| 1 5/16" | 7/8" | — | 7/8" | 7/8" | — | — | — | — |
| 1 3/8" | — | — | — | — | — | — | 1 1/8" | — |
| 1 7/16" | — | 7/8" | — | — | 7/8" | 7/8" | — | 1" |
| 1 1/2" | 1" | — | 1" | 1" | — | — | 1 1/4" | — |
| 1 5/8" | — | 1" | — | — | 1" | 1" | — | 1 1/8" |
| 1 11/16" | — | — | 1 1/8" | 1 1/8" | — | — | — | — |
| 1 13/16" | — | 1 1/8" | — | — | 1 1/8" | 1 1/8" | — | 1 1/4" |
| 1 7/8" | 1 1/4" | — | 1 1/4" | 1 1/4" | — | — | — | — |
| 2" | — | 1 1/4" | — | — | 1 1/4" | 1 1/4" | — | 1 3/8" |
| 2 1/16" | 1 3/8" | — | 1 3/8" | 1 3/8" | — | — | — | — |
| 2 3/16" | — | 1 3/8" | — | — | 1 3/8" | 1 3/8" | — | 1 1/2" |
| 2 1/4" | 1 1/2" | — | 1 1/2" | 1 1/2" | — | — | — | — |
| 2 3/8" | — | 1 1/2" | — | — | 1 1/2" | 1 1/2" | — | — |
| 2 7/16" | 1 5/8" | — | — | — | — | — | — | — |
| 2 9/16" | — | 1 5/8" | — | — | 1 5/8" | 1 5/8" | — | — |
| 2 5/8" | 1 3/4" | — | — | — | — | — | — | — |
| 2 3/4" | — | 1 3/4" | — | — | 1 3/4" | 1 3/4" | — | — |
| 2 13/16" | 1 7/8" | — | — | — | — | — | — | — |
| 2 15/16" | — | 1 7/8" | — | — | 1 7/8" | 1 7/8" | — | — |
| 3" | 2" | — | — | — | — | — | — | — |
| 3 1/8" | — | 2" | — | — | 2" | 2" | — | — |
| 3 3/8" | 2 1/4" | — | — | — | — | — | — | — |
| 3 1/2" | — | 2 1/4" | — | — | 2 1/4" | 2 1/4" | — | — |
| 3 3/4" | 2 1/2" | — | — | — | — | — | — | — |
| 3 7/8" | — | 2 1/2" | — | — | 2 1/2" | 2 1/2" | — | — |
| 4 1/8" | 2 3/4" | — | — | — | — | — | — | — |
| 4 1/4" | — | 2 3/4" | — | — | 2 3/4" | 2 3/4" | — | — |
| 4 1/2" | 3" | — | — | — | — | — | — | — |
| 4 5/8" | — | 3" | — | — | 3" | 3" | — | — |
| 4 7/8" | 3 1/4" | — | — | — | — | — | — | — |
| 5" | — | — | — | — | 3 1/4" | 3 1/4" | — | — |
| 5 1/4" | 3 1/2" | — | — | — | — | — | — | — |
| 5 3/8" | — | — | — | — | 3 1/2" | 3 1/2" | — | — |
| 5 5/8" | 3 3/4" | — | — | — | — | — | — | — |
| 5 3/4" | — | — | — | — | 3 3/4" | 3 3/4" | — | — |
| 6" | 4" | — | — | — | — | — | — | — |
| 6 1/8" | — | — | — | — | 4" | 4" | — | — |

*Regular square only.

# TORQUE CONVERSION

| NEWTON METRES (N•m) | POUND-FEET (LB.-FT.) |
|---|---|
| 1 | 0.7376 |
| 2 | 1.5 |
| 3 | 2.2 |
| 4 | 3.0 |
| 5 | 3.7 |
| 6 | 4.4 |
| 7 | 5.2 |
| 8 | 5.9 |
| 9 | 6.6 |
| 10 | 7.4 |
| 15 | 11.1 |
| 20 | 14.8 |
| 25 | 18.4 |
| 30 | 22.1 |
| 35 | 25.8 |
| 40 | 29.5 |
| 50 | 36.9 |
| 60 | 44.3 |
| 70 | 51.6 |
| 80 | 59.0 |
| 90 | 66.4 |
| 100 | 73.8 |
| 110 | 81.1 |
| 120 | 88.5 |
| 130 | 95.9 |
| 140 | 103.3 |
| 150 | 110.6 |
| 160 | 118.0 |
| 170 | 125.4 |
| 180 | 132.8 |
| 190 | 140.1 |
| 200 | 147.5 |
| 225 | 166.0 |
| 250 | 184.4 |

| POUND-FEET (LB.-FT.) | NEWTON METRES (N•m) |
|---|---|
| 1 | 1.356 |
| 2 | 2.7 |
| 3 | 4.0 |
| 4 | 5.4 |
| 5 | 6.8 |
| 6 | 8.1 |
| 7 | 9.5 |
| 8 | 10.8 |
| 9 | 12.2 |
| 10 | 13.6 |
| 15 | 20.3 |
| 20 | 27.1 |
| 25 | 33.9 |
| 30 | 40.7 |
| 35 | 47.5 |
| 40 | 54.2 |
| 45 | 61.0 |
| 50 | 67.8 |
| 55 | 74.6 |
| 60 | 81.4 |
| 65 | 88.1 |
| 70 | 94.9 |
| 75 | 101.7 |
| 80 | 108.5 |
| 90 | 122.0 |
| 100 | 135.6 |
| 110 | 149.1 |
| 120 | 162.7 |
| 130 | 176.3 |
| 140 | 189.8 |
| 150 | 203.4 |
| 160 | 216.9 |
| 170 | 230.5 |
| 180 | 244.0 |

# SI METRIC-CUSTOMARY CONVERSION TABLE

| Multiply | by | to get equivalent number of: |
|---|---|---|
| **LENGTH** | | |
| Inch | 25.4 | millimetres (mm) |
| Foot | 0.304 8 | metres (m) |
| Yard | 0.914 4 | metres |
| Mile | 1.609 | kilometres (km) |
| **AREA** | | |
| Inch² | 645.2 | millimetres² (mm²) |
| | 6.45 | centimetres² (cm²) |
| Foot² | 0.092 9 | metres² (m²) |
| Yard² | 0.836 1 | metres² |
| **VOLUME** | | |
| Inch³ | 16 387. | mm³ |
| | 16.387 | cm³ |
| | 0.016 4 | litres (l) |
| Quart | 0.946 4 | litres |
| Gallon | 3.785 4 | litres |
| Yard³ | 0.764 6 | metres³ (m³) |
| **MASS** | | |
| Pound | 0.453 6 | kilograms (kg) |
| Ton | 907.18 | kilograms (kg) |
| Ton | 0.907 | tonne (t) |
| **FORCE** | | |
| Kilogram | 9.807 | newtons (N) |
| Ounce | 0.278 0 | newtons |
| Pound | 4.448 | newtons |
| **TEMPERATURE** | | |
| Degree Fahrenheit | (°F-32) ÷ 1.8 | degree Celsius (C) |

°F  -40   32   0   40   80   98.6   120   160   200   212
°C  -40  -20   0   20   37   40   60   80   100

| Multiply | by | to get equivalent number of: |
|---|---|---|
| **ACCELERATION** | | |
| Foot/sec² | 0.304 8 | metre/sec² (m/s²) |
| Inch/sec² | 0.025 4 | metre/sec² |
| **TORQUE** | | |
| Pound-inch | 0.112 98 | newton-metres (N·m) |
| Pound-foot | 1.355 8 | newton-metres |
| **POWER** | | |
| Horsepower | 0.746 | kilowatts (kW) |
| **PRESSURE OR STRESS** | | |
| Inches of mercury | 3.377 | kilopascals (kPa) |
| Pounds/sq. in. | 6.895 | kilopascals |
| **ENERGY OR WORK** | | |
| BTU | 1 055. | joules (J) |
| Foot-pound | 1.355 8 | joules |
| Kilowatt-hour | 3 600 000. or 3.6x10⁶ | joules (J = one W's) |
| **LIGHT** | | |
| Foot candle | 10.764 | lumens/metre² (lm/m²) |
| **FUEL PERFORMANCE** | | |
| Miles/gal | 0.425 1 | kilometres/litre (km/l) |
| Gal/mile | 2.352 7 | litres/kilometre (l/km) |
| **VELOCITY** | | |
| Miles/hour | 1.609 3 | kilometres/hr (km/h) |

# DECIMAL AND METRIC EQUIVALENTS

| Fractions | Decimal In. | Metric MM. | Fractions | Decimal In. | Metric MM. |
|-----------|-------------|------------|-----------|-------------|------------|
| 1/64 | .015625 | .39688 | 33/64 | .515625 | 13.09687 |
| 1/32 | .03125 | .79375 | 17/32 | .53125 | 13.49375 |
| 3/64 | .046875 | 1.19062 | 35/64 | .546875 | 13.89062 |
| 1/16 | .0625 | 1.58750 | 9/16 | .5625 | 14.28750 |
| 5/64 | .078125 | 1.98437 | 37/64 | .578125 | 14.68437 |
| 3/32 | .09375 | 2.38125 | 19/32 | .59375 | 15.08125 |
| 7/64 | .109375 | 2.77812 | 39/64 | .609375 | 15.47812 |
| 1/8 | .125 | 3.1750 | 5/8 | .625 | 15.87500 |
| 9/64 | .140625 | 3.57187 | 41/64 | .640625 | 16.27187 |
| 5/32 | .15625 | 3.96875 | 21/32 | .65625 | 16.66875 |
| 11/64 | .171875 | 4.36562 | 43/64 | .671875 | 17.06562 |
| 3/16 | .1875 | 4.76250 | 11/16 | .6875 | 17.46250 |
| 13/64 | .203125 | 5.15937 | 45/64 | .703125 | 17.85937 |
| 7/32 | .21875 | 5.55625 | 23/32 | .71875 | 18.25625 |
| 15/64 | .234375 | 5.95312 | 47/64 | .734375 | 18.65312 |
| 1/4 | .250 | 6.35000 | 3/4 | .750 | 19.05000 |
| 17/64 | .265625 | 6.74687 | 49/64 | .765625 | 19.44687 |
| 9/32 | .28125 | 7.14375 | 25/32 | .78125 | 19.84375 |
| 19/64 | .296875 | 7.54062 | 51/64 | .796875 | 20.24062 |
| 5/16 | .3125 | 7.93750 | 13/16 | .8125 | 20.63750 |
| 21/64 | .328125 | 8.33437 | 53/64 | .828125 | 21.03437 |
| 11/32 | .34375 | 8.73125 | 27/32 | .84375 | 21.43125 |
| 23/64 | .359375 | 9.12812 | 55/64 | .859375 | 21.82812 |
| 3/8 | .375 | 9.52500 | 7/8 | .875 | 22.22500 |
| 25/64 | .390625 | 9.92187 | 57/64 | .890625 | 22.62187 |
| 13/32 | .40625 | 10.31875 | 29/32 | .90625 | 23.01875 |
| 27/64 | .421875 | 10.71562 | 59/64 | .921875 | 23.41562 |
| 7/16 | .4375 | 11.11250 | 15/16 | .9375 | 23.81250 |
| 29/64 | .453125 | 11.50937 | 61/64 | .953125 | 24.20937 |
| 15/32 | .46875 | 11.90625 | 31/32 | .96875 | 24.60625 |
| 31/64 | .484375 | 12.30312 | 63/64 | .984375 | 25.00312 |
| 1/2 | .500 | 12.70000 | 1 | 1.00 | 25.40000 |

0838

# INDEX